BAD NATURALIST

BAD NATURALIST

One Woman's Ecological Education on a Wild Virginia Mountaintop

PAULA WHYMAN

TIMBER PRESS | PORTLAND, OREGON

Copyright © 2025 by Paula Whyman.
All rights reserved.

This is a creative work depicting events in the life of the author to the best of her recollection. Now and then, time has been compressed or chronology shifted to accommodate the story. Dialogue consistent with the character or nature of the person speaking may have been edited, supplemented, or reconstructed from the author's recollections. Everyone mentioned in the book is an actual person; the names of some people have been changed to respect their privacy or to avoid confusion. The information in this book is true and complete to the best of our knowledge, but all recommendations are made without guarantee on the part of the author or Timber Press.

Illustration on page 6 by Vincent James

Hachette Book Group supports the right to free expression and the value of copyright. The purpose of copyright is to encourage writers and artists to produce the creative works that enrich our culture. The scanning, uploading, and distribution of this book without permission is a theft of the author's intellectual property. If you would like permission to use material from the book (other than for review purposes), please contact permissions@hbgusa.com. Thank you for your support of the author's rights.

Timber Press
Workman Publishing
Hachette Book Group, Inc.
1290 Avenue of the Americas
New York, New York 10104
timberpress.com

Timber Press is an imprint of Workman Publishing, a division of Hachette Book Group, Inc. The Timber Press name and logo are registered trademarks of Hachette Book Group, Inc.

Printed in the USA on responsibly sourced paper
Text and cover design by Vincent James
The publisher is not responsible for websites (or their content) that are not owned by the publisher.

ISBN 978-1-64326-217-8
A catalog record for this book is available from the Library of Congress.

FOR BILL

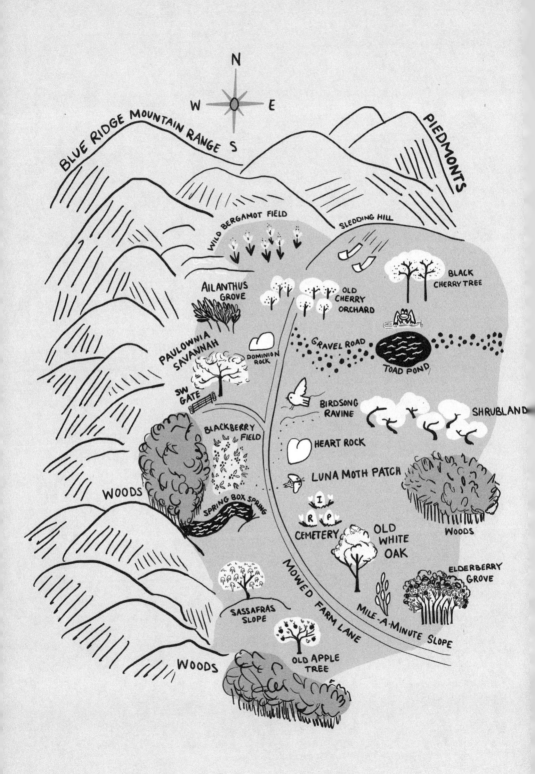

The secret of life is to have a task, something you devote your entire life to, something you bring everything to, every minute of the day for the rest of your life. And the most important thing is, it must be something you cannot possibly do.

—Henry Moore

How could a weed be a book?

—Aldo Leopold

CONTENTS

Preface 11

Part I: The Mountain of Questions

Chapter 1. The Terrarium 18
... in which I'm stung by wasps and sick over sea lions, and I finally find the perfect picnic spot

Chapter 2. Finding the Way 31
... in which I fail at navigating and at gardening

Chapter 3. A Not-Gardener's Education 41
... in which I'm introduced to a native prairie, a plant whisperer, and a no-holds-barred approach to meadow restoration

Chapter 4. Making a Plan 52
... in which I ask the government for help, meet a bee-hugger, and learn that "baseline condition" is in the eye of the beholder

Chapter 5. Once Upon a Time, Five Hundred Million Years Ago 70
... in which mountains rise and erode, apples come and go, and I question the meaning of land

Part II: The Mountain of Weeds

Chapter 6. Two Ways of Looking at a Hillside 88
... in which a farmer's weed is my wildflower

Chapter 7. We Brought This on Ourselves *102*
... in which a wildflower becomes a weed, weeds threaten to engulf the planet, and the original invasive species is...me

Chapter 8. Unintended Consequences Farm *116*
... in which an invasive vine eats my brain

Chapter 9. Oh, Deer *131*
... in which weeds follow the herd

Chapter 10. The Bad Tree *147*
... in which a tree escapes a garden and conquers the world

Part III: The Mountain of Hope

Chapter 11. On Fire *166*
... in which fire prevents fire, native plants want to be burned, and I wait for the wind to die down

Chapter 12. The Mountain and the Vole Hill *186*
... in which an ant guards a tree, a lichen becomes a bird's nest, and a vole plants a seed

Chapter 13. Blackberry Fields Forever *207*
... in which there can be too much of a good thing

Chapter 14. Let it Grow *219*
... in which good things come to those who wait

Chapter 15. Steward of Secrets *236*
... in which I don't know all the answers

Suggested Reading *239*
Acknowledgments *250*

PREFACE

When I was a kid and I wanted to be alone and quiet, I'd walk into the woods behind my elementary school, far enough that I could no longer see the playground or the baseball field or the apartments through the trees. I stopped walking before I reached a large cemetery's maintenance area, where I'd once discovered piled-up concrete grave liners and an eerie silence. I was satisfied when I felt surrounded by trees, and I could pretend to be far away. The scent was green, it was rotting leaves, and sometimes it was the sharp, carcassy smell of skunk spray. My destination was a shallow stream about three feet across, not much deeper than the puddling on my street from an overwhelmed storm drain after a quick heavy rain. It was like something out of a story I'd once read: two smooth-barked trees on opposite sides of the stream reaching toward each other, their branches coming together to form an arc over the water. The water was clear, and I could see the bottom. Kids would claim to catch crayfish in this stream, but I was doubtful. That didn't stop me from looking; I'd sit on the same fallen log and stare into the water, hoping I could report seeing a crayfish, or any fish. Usually all I saw were beetles.

Around that time, I tore through James Herriot's books about his life as a veterinarian, and I decided I would be a veterinarian, too. That was the only job I knew of where I could take care of animals all the time, and I was sure I could do it in a picturesque countryside like Herriot's Yorkshire Dales. But the more I read about veterinary medicine, the more I realized that along with the saving and curing, veterinary work also involved a lot of dying. I didn't want to be responsible for any creature dying, or for killing anything, even when it was the humane thing to do. I gave up on becoming a veterinarian, and became a vegetarian instead (right as my mother was putting a chicken dinner on the table). Eventually, I gave that up as well and settled on becoming a writer.

I came to this mountain after thirty years of searching for a place in the country, during which I mostly didn't believe we'd end up finding a place at all. Most of the search was about standing in a spot on a river or in a hayfield or on a hilltop and daydreaming. The daydreaming was satisfying in itself for a long time. To actually move forward with a plan seemed unrealistic. For years, I didn't want the added responsibility I knew would come with it, and that wasn't even about the land—I was raising kids; I had no spare time. And I didn't understand land ownership. What did it mean, to *have* it? I'm still not sure that I know.

Eventually of course, children grow; mine did, and they can take care of themselves now. A decision was made at dinner, after once again hiking and driving around the hilly county whose natural beauty and rural life we'd always appreciated. When I was in this place, I realized, I wasn't on edge. The stress I'd long carried fell away. As an adult, I still needed a place like those woods, where I could be quiet. We restarted our search in earnest, after a long hiatus, and I was driven by the particular goal of establishing a native meadow wherever we ended up—a neat, organized, narrowly defined project. I don't think I realized until I was in the middle of it that I was ready for a new kind of responsibility; I'll call it stewardship. To finally own this land was a decision long in the making, even if it may have seemed, at the time it finally happened,

Preface

impulsive, chaotic, irrational. It was all of those things, and none of those things.

Sometimes a book or story idea will emerge when I overhear a line of dialogue; the words float around in my head for a while and then the line sort of falls into a slot and I can see where it might fit, and the larger story grows up around those initial words. When I'm done I may not remember the piece of dialogue that started the whole thing; maybe it's not even in the story anymore. In this case, my idea to plant a meadow was like the bit of dialogue that had yet to be developed. It led me to the big idea, when, standing on the mountain for the first time, I was overcome by the epic potential of this place.

I started the project of restoring the land in my head, researching, before I started it in reality, and I was months into it in reality before I decided to write a book about it. I was working on a novel at the time. This project so successfully took over my mind, I don't remember what that novel was going to be about. I realized then, there was no question that I'd write about the mountain and what I'm trying to do here. It would have been impossible to write anything else, and it would have been impossible not to write at all. I'm driven to it, even if what I end up accomplishing is inherently ephemeral, and, like everything else, I can't be sure of what will happen to it after I'm gone.

As with other projects I've attempted, I may not know quite what I'm doing, but I know that I'll either figure it out, or I'll figure out that it doesn't work. Like many writers, I'll pursue an idea, I'll go down that rabbit hole, because that's the only way I can find out if it can be made real. Often the rabbit hole turns out to be no more than a divot.

In an interview in *Outside* magazine, documentarian Ken Burns described his many, varied project ideas as akin to the numbered balls in a lottery machine. They bounce around in your head, "and then one drops down to your heart." When I came to this mountain, what finally landed in my heart was a much larger ambition than planting a small meadow. I thought I could make a difference here. That seems like such an earnest and clichéd desire to admit. Whenever I feel dangerously earnest I immediately temper that feeling with some good self-mockery. I'm reminded of a line from a sitcom spoken by a megalomaniacal tech company founder: "I don't want to live in a world where someone else makes the world a better place better than [I] do!"

One of my kids told me, "Being able to make a difference is a privilege." He's right. There have been many times in my life when I've been completely occupied with simply getting by. Now, it's a privilege to be able to do what I'm doing, hiking around a mountain trying to figure out why there are so many lichens and voles and thistles and thorns and how some grasses can be brown even when they're not dead. And writing about it—the meadow and the mountain and what I'm doing here—at the same time I'm trying to understand all of it. The result is me wandering and stumbling and gradually getting to know this place, the plants and creatures and how they interact. My hope is that my experience will resonate with anyone who is interested in getting to know any part of the natural world they encounter in their day-to-day lives.

I was hit by lightning while working on this book, and I wasn't even outdoors. It was a secondary bolt that shot through the floor and the touchpad on my laptop, zipped up my arm, and knocked me out of my chair. I like to think it helped my process. For a few hours after, my hand was tingling. I would be struck, metaphorically, many times as I tried to "fix" the meadows here, surprised by what I learned about the way the living things rely on each other, startled by the paradoxes I encountered that are inherent in attempts to practice conservation, all as I struggled with questions to which there seemed to be no single "right" answer.

One thing I learned: restoration, as an end point where everything stops, is not possible, not truly. If you somehow, against the odds, can guide a natural place to a perfect moment, exactly what you were aiming for—after reading my story, I think you'll laugh and maybe also cry at the idea—you might be able to keep it that way, briefly. But we're talking about nature, by definition ever evolving, ever changing. I can't control

Preface

the climate, the wind, where plants decide to grow, the creatures that carry seeds on their feathers and fur and in their droppings and leave them where I don't want them, the neighbor who likes those pretty invasive trees with the big purple flowers, the nurseries that sell nonnative plants that escape gardens and end up here, the meandering stream that distributes weed seeds along its banks. I can't keep the land one way forever. I could drive myself batty trying. This story is about how I tried, and how I both failed and succeeded, and how the true benefit, as with so many efforts, lay not only in how the plants grew but how I grew in the process.

Back when I started, I couldn't live on the mountain while I studied and worked on the land. The road up was barely passable, there was no source for drinking water, no power, and no shelter of any kind, besides a tent that kept blowing away and a broken-down hunting stand deep in the woods, long since stove in by fallen branches and inhabitable only by field mice. Much of what I describe in this book took place under those circumstances. Now I finally have water and power and a place to live here (and an actual working bathroom, which it turns out is far more pleasant than squatting behind a boulder at the exact moment a visitor shows up, calling your name from the other side of said boulder as you frantically pull up your pants).

There was, and still is, so much I don't know about this place and the life here, and that's not bad. In sharing my early lack of knowledge and understanding, my many mistakes and failures, and a few successes, whether intentional or accidental, I hope someone out there might also come to notice the plants around them, to see all the beings—plants, birds, insects, wildlife—that live in their own backyard or farm or city park in a different light. For someone who was fascinated by all kinds of wildlife, but who didn't spend much time paying attention to plants before, I now think about them a lot more than I ever imagined I would.

As I was inspired by Isabella Tree's book *Wilding*, and her work restoring her farm in Sussex, England, I hope that someone will read this book and consider what may be possible, that it might lead them to think more and differently about the landscape, wherever it is they find themselves, to wonder what might have made a place what it is, and what it might be capable of becoming. I think about my impact on the land in a new way, and when I say *the land*, I mean all of it, not only the spot where I happen to be standing. Maybe this will happen to you, too.

PART I: THE MOUNTAIN OF QUESTIONS

Chapter 1

THE TERRARIUM

... in which I'm stung by wasps and sick over sea lions, and I finally find the perfect picnic spot

My first time on the mountain, it was August; relentless heat, bright sun beating down, no place to hide. I was surrounded by the frenzied growth of a meadow that had been left to its own devices for years: endless fields of yellow wildflowers taller than I was, bumble bees zigzagging past my head, butterflies fluttering about, and ticks climbing up my socks. I was bombarded with impressions, and, with little context for interpreting or understanding what I was experiencing, it was all I could do to take it in. I hiked along a narrow path that was mowed only days before to allow space to walk, and within a few minutes I felt like I was submerged in a sea of plants, invisible. I stumbled on a vine stretched like a trip wire across the path and sidestepped coyote dung, a fancy maneuver that would eventually become second nature. I reached the highest spot on the mountain and turned in circles, every angle its own scenic vista. The Blue Ridge was

The Terrarium

laid out like slices of layer cake across the west and southwest horizon, the foothills of the Virginia Piedmont to the south and east.

I've hiked on countless trails in countless places, from teenage misadventures in woods where No Trespassing signs clearly should have been heeded, to moonlit rainforest, to jagged lava field, to the other end of the spectrum—timber-lined mulched trails where I could still hear the traffic on the interstate, and where a public restroom (sometimes clean) waited at the end. But this is different, hiking where I'm the only human I can reasonably expect to see simply because this is now "my" place. But I don't think of it as mine as much as a place I'm now sharing with the creatures that were already here. It's the realm of the bee and the beetle, the raptor and the songbird, the snake—venomous and non—the frog and the coyote, the turtle, the tick, and the turkey. And it's the realm of a thousand plants, both desirable and not (depending on your point of view), many of which grew here by chance, brought by these same animals and birds, or brought here because of choices made by humans who were here a long time before me, and who knew more and less than I do and had disparate goals. Those people undoubtedly believed they were making the right choices at the time, based on what they knew. Will future inhabitants look back at what I do here with dismay and incomprehension, with relief, or with indifference?

I'd come here in part because I wanted to plant a meadow. I thought planting a meadow was a way to save native wildlife, which I'd wanted to do, and tried to do in small ways, my whole life. When I was a kid, on summer trips to Rehoboth Beach, I tried to rescue horseshoe crabs that washed up on the sand by transferring them to the bathtub in our hotel room, as if I could create a new home for them there. I'd failed at that, the same way I'd failed to save my pet praying mantis from being run over by a lawn mower. As an adult, I'd tried to save a fledgling robin from broiling on my suburban driveway during an unusual spring hot spell. I watched the bird all morning, knowing that I shouldn't interfere, but when it keeled over in the heat, I laid it carefully in a shoebox and brought it to an animal shelter. It was dead before I arrived.

And then there were the frogs. More than a decade before I knew this mountain existed, I purchased a terrarium and tadpoles for my kids. We watched with anticipation as the tadpoles grew into tiny grass-green frogs. I went to the local pet store and brought home batches of live crickets to feed them. The crickets were not very smart: the moment I released them into the terrarium, they hopped from the fake lawn into the tiny pool of water and drowned. The frogs, for their part, would only eat live crickets. And, for some reason, the frogs didn't hang out in the water as much as I'd expected. Maybe all those dead floating crickets? I kept adding crickets; the crickets kept drowning. Even though I provided plenty of food, it wasn't long before the frogs died, too. One at a time, over a period of weeks, I woke in the morning to discover their stiff little desiccated bodies. I never figured out exactly why they died. Was it the limited food supply? The dry air? Disease? Was it a loud household of people, vacuum cleaners, trombones and trumpets, leaf blowers outside the window, plus the inept person who didn't know how to keep crickets from drowning? I could only guess. They were trapped in the terrarium, restricted to the nourishment and atmosphere I could provide. Whatever the cause, the frogs had gone extinct in their environment, and I'd played a role. Once they were gone, nothing came along and replaced them, because there was no nearby habitat from which other frogs could migrate to fill the gap. I wasn't about to buy more tadpoles and risk killing them again; it was too heartbreaking for everyone.

Not long after, we came into some hermit crabs left by a friend who was moving away. One of my kids asked, if we don't kill the hermit crabs, *then* can we get a dog?

Am I just terrible at saving creatures? Would I be any better at saving places?

Now it's my turn to decide what will happen on this mountain. What's at stake? Only a mountaintop. Only all the living things that call it home. What if I screw it up? Given my track record, there's a strong likelihood that despite my best intentions, I'm going to screw it up.

The Terrarium

* * *

Around twenty-five years ago, on an educational trip led by a paleoclimatologist and a conservation biologist, I was snorkeling in the Pacific Ocean off Darwin Island in the Galápagos while endangered hammerhead sharks swam fifteen feet below me. The sharks ignored me, in the way that most of the creatures in the Galápagos ignored us humans or seemed to see us as a source of amusement. During the course of this trip, a blue-footed booby untied my shoes, a mockingbird stole and hid a fellow traveler's hearing aids, and, more than once, we shared our deck with opportunistic and intransigent sea lions that, along with the local penguins, liked to play with us in the water, including dive-bombing out of nowhere to snout-tap us on our snorkel masks, which is, I guess, the sea lion equivalent of the fist bump. We spent ten days on a small boat, and between lectures about Darwin's finches and the Humboldt current, nearly everyone in our traveling group of twelve became seasick at least once, most of them on the night we crossed the equator. That night, only a stalwart few appeared at dinner where even the boat-wood chairs, heavy as concrete, were rocked away from the table by rough seas. I was (far more humiliatingly) land-sick instead. What did me in was a place called Sea Lion Beach. That may not be the official name, but it should be. It's a wide beach covered with flat boulders, and every boulder is draped with sea lions, none of which even twitched a whisker as we approached. I've not experienced before or since as intense an astringent odor, the odor of sea-lion pee in overwhelming volume; not even bat guano matches it. That and what my confused vestibular system imagined as the rocking of the shore sent me to lose my lunch behind a (sea-lion-free) rock. This wasn't the sea lions' fault. My only truly negative contact with wildlife came courtesy of an invasive species, when I was stung by nonnative wasps massing around a hanger of bananas on the boat.

    It was in the Galápagos islands that I was confronted with the impact of invasive species and habitat loss up close, a harm considerably worse than a swollen arm. I saw the damage caused by feral pigs and goats that reproduce readily and will eat nearly anything. They had been introduced to these islands, and many others, hundreds of years earlier by whalers who seeded islands along their whaling route with a reliable food source. The animals multiplied and munched out on land where they had no natural enemies and where they didn't belong in the first place. Their presence upset delicate island ecosystems; they crushed, ate, and uprooted native plants, and devoured the eggs and trampled the nests of native ground-nesting birds and tortoises.

In the water, overfishing along with unusual current temperatures caused by climate change led to declines in sea life. Even in an extremely remote place like the Galápagos, where so much is protected, there were those who tried to take advantage: the mayor of one of the islands' few towns was caught hoarding off-limits sea cucumbers. Just outside the protected area that creates an invisible boundary around the islands, sharks were harvested for their fins and tossed back into the water to die. These injuries to the islands embedded themselves in my mind like an everlasting painful sting.

What was happening in the Galápagos turned out to be a microcosm of what's happening all over the world. The biologist who led that trip introduced me to the theory of island biogeography, originated by Edward O. Wilson and Robert MacArthur. Back in the 1960s, MacArthur and Wilson noticed that species living on islands in the middle of an ocean are more prone to extinction than those living on larger territories, on larger islands, or on islands that are closer to the mainland. They examined why that was the case and came up with ways to help predict extinctions in such places. But they didn't restrict their discussion to literal islands in the middle of an ocean, like the Galápagos; they applied their theory to isolated ecosystems everywhere, including lands that may as well be islands—those that were once large, continuous wild places, until they were broken up by agriculture or development into smaller, disconnected bits, a situation known as *habitat fragmentation*. These disconnected bits, they contended, behave very much like islands;

The Terrarium

they're subject to similar pressures that decrease biodiversity—the variety of species that live in a given place—by threatening the existence of individual species and whole ecosystems, the interconnecting community of organisms trying to survive there. And, because development was unlikely to end, they predicted the problem would grow worse with time. (Spoiler: it has.)

It seems like just yesterday I was reading Edward O. Wilson's *Diversity of Life* to one of my kids as a bedtime story. (My son may have been the first preschooler to learn about the volcanic destruction of Krakatoa; I think he's forgiven me.) And then, a few years ago, my kids were launched into the world of college and beyond. As a midlife empty nester, my time became more flexible, and I was determined to more actively pursue my longtime avocation, by rescuing a small corner of nature that needed help. I figured I knew more now and could do a lot better than dragging unsuspecting horseshoe crabs off the beach and pouring table salt into bathwater as a substitute for the ocean. I'd recently read Isabella Tree's book *Wilding*, about her years-long project of returning her thirty-five-hundred-acre farm in Sussex, England, to a more authentic wild state. It was the most exciting conservation project I'd heard of in years; I couldn't get it out of my mind. How amazing, how rewarding that must be, to make a difference in a place that large—to rescue not only one creature or one plant, but whole communities of creatures and plants. What if *I* could take a piece of land and restore it to a more natural condition? What if I could create a home for birds and other creatures that are losing more habitat every day? I'd researched it, and I *thought* I knew what it took to plant and maintain a meadow. What if I could do even a small fraction of what Isabella Tree had done? Wouldn't that be something? One of the first things she did was plant a meadow. I could do that—I'd take two or three acres, four at the most, and create a native meadow with wildflowers to attract pollinating insects and local wildlife.

The question was how, and where, would I come up with these acres? My longtime home in the Washington, D.C., suburbs had a standard-size lawn that was regularly trampled down to dirt by our standard-size poodle and her daily zoomies. My husband, B., had planted and landscaped wherever the trampling was less energetic, but there wasn't room for much of a meadow.

As it happened, my desire to plant a meadow coincided with the escalation of our long-running search for retired farmland in the countryside, where we planned to spend the next phase of our lives. After years of being limited to sporadic country visits, we were more than ready for a big change. We wanted nature at its most, well, natural; an outdoor life, where B., an Eagle Scout (once an Eagle Scout, always an Eagle Scout), could go on hikes and unwind by a campfire, and where I could grow vegetables and continue to find new creatures to study—and now, I hoped, plant a meadow.

I pictured an old hayfield, a broken-down barn, an old farmhouse that needed fixing up. I pictured a writing studio in the middle of the meadow I was going to plant, where I'd watch flowers shift in the breeze while awaiting inspiration. But our quest so far had proved inconclusive. Like the Berenstain Bears' search for their perfect picnic spot, from one of my favorite children's stories, I thought we'd never find the right place. My dream meadow plan was destined to remain a dream.

And then one day, here we were, standing on top of a mountain. This was it, and like so many "perfect" places, it turned out to be nothing like what I'd pictured. It was, for sure, an old cow pasture that had not seen cows in years, an old hayfield that had gone to seed. But it was far more than that. (And far less—there was no old barn or old farmhouse for me to fix up; there were no buildings at all, broken down or otherwise.) This was more than a few acres. It was over two hundred acres on a mountain, more than a third of it rolling open meadows—the former cow pasture—some of it quite steep, surrounded by even steeper sloping forests that spilled down the hillsides. A place where hawks circled, voles tunneled, deer wandered, and bears and bobcats lurked, where trees and grasses were bent by wind, where the sky changed every moment; a landscape of preternatural beauty and drama—and a massive tangle of weeds.

The Terrarium

Perfection was too much to expect, right? This hoped-for country paradise turned out to be full of what I would eventually learn were invasive plants. For me, it was still a paradise. If I'm honest, I'll say I'm most alive when I'm trying to fix something, trying to solve a problem. (Why does that have to be true? Can't I sit back and enjoy things as they are? Evidently not.)

The more I learned, the better I could imagine the potential of this place, and the more I felt committed to repairing it. And, the more I learned, the more I had to face that, in this job I'd volunteered myself for, total control was impossible and my definition of success would be, necessarily, elastic.

* * *

There's an inspirational poster that hangs on the wall in my dad's home office. It's been there since I was nine years old. The words on the poster begin, "Nothing in the world can take the place of persistence ... talent will not ... genius will not." (It wasn't until years later that I learned it was a quote from Calvin Coolidge.) Growing up, I found this philosophy exasperating. Couldn't I be allowed to give up sometimes without feeling bad about it? Well, no. I'd be called into my dad's office for a lecture and to stare at that annoying poster. Maybe this is why my tastes as an adult lean more toward parodies of inspirational posters, like the one that depicts an athlete lying on his back on a playing field, covering his face with his hands. It reads "Failure is not an option. It's your destiny."

But my parents have lived by the terms of that persistence directive. Nowhere was it more evident than in the way they tried to help my younger brother. He was born with Williams syndrome, a genetic disorder that created obstacles for him in every aspect of his life, physical, emotional, and intellectual. I grew up watching my parents work devotedly, trying the best they could to equip him for life in the mainstream world, even though I think we all knew (but never explicitly said) that an independent life would never be truly possible for him. Maybe those rose-colored glasses were what helped my parents to persist, to my brother's very real benefit. They made sure he became the most capable and contented person he could be, within the limits of his disabilities.

I have, for better or worse, inherited my parents' tendency to persist. Nature? Nurture? A little of both, probably. I haven't contended with anything nearly as serious as my brother's problems; in comparison, mine have been no more than minor inconveniences. But we all have our private disappointments, and as I get older, there have been times when I've felt increasingly boxed in by health concerns, as if the world is narrowing around me, and limits are something I've never been good at accepting. I wonder why? My father, at age ninety-one, fully cerebral, thankfully, but with extremely poor balance, decided to stand on the sofa to hang a sign on the wall while my mother was in the shower. Before she got in the shower, my mother told him not to stand on the sofa. Why would she even think of telling him this? Only because she's lived with him for sixty years and has seen him take similar risks that ended in near-serious injury, including the time he fell through the ceiling of the garage after trying to balance on attic joists, and the time he cut off his finger replacing the fan belt in the car. This time, he lost his balance on the mushy sofa cushions and fell and injured his neck. The doctor said he was lucky he wasn't paralyzed. Now, at ninety-three, after months of rehab, he can walk, haltingly, almost normally, without assistance. The other sign on my dad's office wall says "The salesman's job starts when the customer says no."

My limits have meant I've had to forgo my favorite outdoor exercise and my favorite foods. I'm a runner who can no longer run, and a baker who can no longer eat bread. I've had to cut out caffeine and wine and dairy, too, part of a long list of culinary deprivations due to a GI disorder. This mountain may have good bones, but I don't; I have osteoporosis. I have, on different occasions, broken a shoulder, an elbow, and a kneecap, plus three toes. (My doctor tells me the toes don't count.) I tore cartilage in both of my hips because of a malformation I only learned about after three years of what would have been ill-advised yoga classes. I'm also a

The Terrarium

klutz, if that's not already clear. I typed this chapter while wearing a brace on a sprained hand, because I tripped over the leg of a picnic bench that had been sitting in the same spot on my own patio for twenty years. I finished this book on crutches, after finally undergoing surgery to replace the torn cartilage in one of my hips. I keep asking the doctor when I'll be allowed to run again. Where is the line between being persistent and plain stubborn?

My physical condition in some ways mirrors the condition of this place—I appear to be in decent shape, but don't ask for an MRI. When I first wandered this meadow, it looked pristine to me. But now that I've learned to identify the many species of invasive plants, I can't unsee them. I worry I won't be able to stop them from destroying an ecosystem that's already struggling. This land was shaped by hundreds of years of human interference, and it will require a different kind of human interference in order to thrive.

* * *

More than half of U.S. forestland is in private hands, and most of the grasslands in the United States—85 percent—belong to private landowners. Grasslands are the most endangered ecosystems in the world. But what do I mean when I say *grassland* and *meadow*? A meadow isn't automatically a grassland—it depends on the balance of what grows there, and to some degree its history, as well as on who's defining it. The meadows here, I'd eventually learn, are a mix of native and nonnative grasses, wildflowers (also known as forbs), and invasive plants. Fields with more grasses can be considered grasslands, and those with more forbs are meadows. That variation can be a good thing, because it leads to greater biodiversity—the variety of species that make up the mountaintop community—even if the abundance of nonnative plants here may not be a good thing.

In the Southeast, grasslands have declined by 90 percent. In Virginia, where this stunning, scraggly mountaintop is located, the Piedmont prairie, the native grassland community that once dominated the region east of the Blue Ridge mountains, is considered extinct; only small remnants hidden in tiny pockets around the state survive. With those numbers in mind, you can begin to understand why conservation of privately owned forests and grasslands is critical.

Back in 1962, even before MacArthur and Wilson had published *The Theory of Island Biogeography*, an independent researcher named Frank Preston declared that "it is not possible to preserve in a state or national park a complete replica on a small scale of the fauna and flora of a much larger area." Parks alone can't begin to help us ward off the worst effects of climate change, human development, the influx of invasive species, and other pressures that endanger biodiversity. Even if we put all the public lands together—all the national, state, and local parks and nature preserves—without the participation of private landowners, our parks would end up becoming species museums showcasing a handful of plants and animals, those lucky enough to survive the isolation. My interest in nature and ecology has often led me to deep dives into endangered ecosystems like coral reefs, mangrove swamps, and cloud forests, but I'd heard little about the threats to grasslands and their inhabitants; nor was I fully aware of the role I could play, as one person, in helping to save a native ecosystem.

Standing on the mountain, with all those acres of rolling hills unfolding in front of me, my goal to plant a small patch of meadow seemed timid. This land was big—I should think bigger! What if I could return this mountaintop to its natural glory? It would serve as a living example of how to restore native meadows! Pollinators would come from all around! I pictured sheep grazing on one of the hillsides. Just a handful of sheep. I'd make sheep-milk cheese. (I can't eat dairy. I was clearly losing my mind.) I'd put up a fence to protect the sheep from the coyotes and bobcats and bears. Better make that an electric fence. Already I was taming the wilderness.

Even as I dreamed up a Percy-Shelleyesque vision of myself communing with nature, skipping through meadows, bluebirds winging

The Terrarium

around my head, I was hit with second thoughts. I had the nagging sense that I might be glossing over a few important factors (besides poison ivy and rattlesnakes). Like, at this stage in my life, and considering my various physical limitations, did I really want to take on such a big, ambitious project with an open-ended timeline and a learning curve as steep as one of those hills? Hadn't I reached what novelist Richard Ford had once dubbed the "permanent period," when I'm supposed to be settled into a long (hopefully) pleasant plateau? Wasn't the point that, since my child-rearing days were over, I could finally eschew at least some burden of long-term responsibility and revel in the freedom of empty-nesthood? Shouldn't I be taking a long, well-earned break, resting on laurels or on a lounge chair somewhere with a drink with a tiny umbrella in it? But just as a vacation with no time set aside for writing never seemed like a real vacation to me, taking a break has never felt like a real break to me.

I understood, in theory, that caring for land was a major commitment—I would soon understand it even better in practice. To complicate matters, not only was there no house to live in up here, not even a shed for storing tools, there was no power, and no working well. This land would require time, sweat, single-mindedness: your basic obsession. In other words, exactly my jam.

So I chose to try to restore these two hundred acres, to attempt to transform ailing fields into native meadows, and a barren forest floor into a teeming native understory. Even though, outside of the knowledge I'd gleaned from books like *Wilding* and "weed warrior" expeditions with my kids along the Potomac River, I had no real idea what would be involved in doing this work on a vast mountaintop, or what the end point of such a project would look like—much less whether a true end point was possible.

If this course of action seems hopeful, foolish, and delusional, that's the wheel around which my feelings about it cycle, sometimes in a single day. My frustrating limitations aside, what I do have going for me is a tendency to become preoccupied with a topic, studying it and talking about it nonstop (much to the irritation of my family). At various points in my life, I'd done this with mangrove trees, gray whales, carpenter bees, rats, sea urchins, carpet beetles, lizards, and flying squirrels. The mountain was not only a subject, but a place I could dive into and lose myself, a source of fascinating and unlimited information to fill my brain and give me a sense of purpose. And unlike most of the places that inspired those other fixations, the mountain wasn't a place I would only visit—it would actually be my home. I'd literally live with the outcome, the success or failure of my endeavors. I didn't know yet how that would feel or how it would influence my decisions.

* * *

It helps to remember those poor dead frogs in terms of island biogeography. An island can be any isolated place, from an actual island, like Fernandina in the Galápagos, to my terrarium. If we shrink a creature's habitat down to an artificially small size, where there is limited food and no accessible path to another viable habitat, it puts tremendous pressure on every form of life there. An island can be a forest surrounded by a housing development; a hedgerow bordering a tilled field; it can be a single plant, if you're a beetle and you spend your entire life cycle on one plant; it can be a backyard butterfly garden, or a highway median; or it can be a meadow surrounded by woods on top of a mountain.

My belief that I can revitalize this place has an almost magical quality to it, as if, as the land grows healthier, I'll grow stronger, too. This may be magical thinking, but it's also an exercise in hope. Standing on top of the mountain, contemplating the idea of somehow bringing change to this formidable landscape, is a dizzying experience. Whatever happens, like the wildflower seeds that stick to my trousers, the land grabs hold of me, and it won't let go.

Chapter 2

FINDING THE WAY

... in which I fail at navigating and at gardening

The first home I can remember was a tiny brick and siding split-level on a dead-end street in a quiet suburb of Washington, D.C. The streets were narrow, backyards were small, and the houses were close together. We lived on a corner, where the lawn wrapped around and created what felt to me then like a large front yard. Our lawn was full of dandelions, and I was discouraged from blowing on the seeds because it would spread them, so instead I made their heads pop off, which kept them from seeding and was also fun. There was clover, good for searching for the lucky ones, which I never found. Other than the lawn, many of the plants in our yard were exotic, which at the time was seen as something special. These plants seemed to fit with the previous owner's claim that she was related to a famous person. She left a few pieces of similarly exotic furniture that she insisted had once belonged to her cousin, Ernest Hemingway. (If this is true, Hemingway's table is now my father's workbench in my parents' garage.)

We were particularly proud of the Chinese silvergrass that grew in a giant tuft at the bend where the front yard became the side yard. And next to the backyard gate was a mimosa tree that reached the second-story window and exploded with fluffy pink flowers in spring. My mother took pride in the unusual plants in our yard, even when the mimosa dropped its seed pods all over the lawn. Red roses grew along the backyard fence, from which I'd spy on our neighbor Mr. M, as if something fascinating might happen while he pushed his reel mower back and forth on his small patch of lawn. My mother loved the roses, but she eventually dug them up because our German shepherd, Rex, kept eating them, a habit that could prove fatal.

Now, many of the plants that were once prized as exotic, like the Chinese silvergrass and the mimosa, are known to be invasive—prolific, introduced plants that can escape from gardens and push out native plants, threatening whole ecosystems on land and in the water. Aided by human disturbance, the spread of such plants has become an urgent threat to biodiversity worldwide. But back when I was growing up in the 1970s, these plants were something to be admired, a feature that distinguished our otherwise unremarkable house.

There was one tree in our yard that was native—not that I knew it then—the pussy willow. It was my favorite. I've yet to see another one of these trees anywhere. Years later, when I was picking up one of my kids at school, I noticed the secretary was keeping pussy willow branches in a dry vase on her desk. I admired them and told her about the tree at the house where I was born. When I stopped in to return my pass, she presented me with a wrapped bundle of the branches.

"Take these home," she said. "You don't need to do anything; they'll keep forever." That was more than fifteen years ago, and they're still in the same condition as when she gave them to me. When I look at them, I return to the tree at the side of the little brick house where I used to wait

Finding the Way

impatiently for spring so I could run my fingers over the soft fuzzy catkins and imagine they were tiny paws.

There's a map of that yard in my head. It's more precise than the map of the inside of the house. The front yard was where we would sit on folding chairs and watch the July 4th fireworks that were launched from the parking lot of a nearby shopping center. The back patio was where I "saved" one hundred live cicadas in a butterfly house. The front stoop was where the milk was delivered in bottles, and the front walk was where I tried roller-skating for the first time. The narrow stretch of grass between our house and the house next door was where I sat under the willow with my best friend, trying to crack open rocks, sure I'd find a geode inside.

* * *

Before I can think about changing anything on the mountain, I need to get to know it, to develop my own internal map of the place. I have a satellite map showing the shape and location of forests and meadows, and I have a plat, a surveyor's rendering of the farm's boundaries and corners, marked with elevations and topo lines from the U.S. Geological Survey. The topo shows me that the whole farm sits above nine hundred feet, and where the lines are closest together, the slopes are steepest. But I don't need a map to tell me that; my knees tell me well enough. I can see where the ridges are, because the lines curve against each other like back-to-back camel humps. The plat tells me that most of the corners are marked with iron rods sunk into the ground. One was marked by a ribbon on an oak tree, but that oak has since fallen. There are thirty-one of these corners. In other words, this land is nowhere near a regular rectangle, or even a parallelogram, like the plats I'm used to seeing. The boundaries form a shape resembling an awkward blocky drawing of a hopping rabbit. It's a shape that owes itself to the way previous owners sold off pieces of the slopes going all the way around the mountain. It's a reminder that the land's current condition also owes itself to the work of the people who were here before me, what they touched and what they left alone.

In the beginning, the mountain felt like a giant mystery to me, and in many ways it still does. My way through mysteries to understanding has always been to write things down. So, I drew up a map. It wouldn't be much use to anyone else, since it doesn't mark the way so much as highlight places I use as landmarks to help orient myself. I sketched in the pages of an old composition book I'd managed to hold on to since elementary school. The book was empty, other than my name inscribed in red pen on the cover in someone else's handwriting. I couldn't help wondering if there was an assignment way back then that I'd failed to complete, since there was a time when I was known to shove all of my unfinished work, what we used to call "dittos," into my lunchbox and throw it away when I got home. It seemed time to fill the notebook with something I wanted to learn; it felt like the right place to record my first walks on the mountain.

I didn't know the names of most of the plants here yet, so I described them with a doodle and a note: "purplish stem" or "wispy seed pods" or "strange sap." I'd skipped out on most of a college botany course, a move I regretted at exam time and even more now. Although I'd spent decades reading about wildife or insects or theories of island biogeography, I paid less attention to plants. At first, I only knew those that had caught my eye for specific reasons, like the mangrove trees on Sanibel Island, where I used to pull up a kayak and watch tiny crabs skitter among the gangly exposed roots. The names of many wildflowers simply never stuck, and because I'd never been a gardener, never planted a full-fledged garden myself, the names of even some of the more common garden plants can still escape me. I knew what I thought of as the "main" trees—oak and maple and poplar and willow—those that grew in places where I'd lived. But in general, and despite passing a course as part of Virginia Tech's forest landowner education program, my tree identification skills were pretty shaky, and as I moved through the meadows, I fumbled through a stack of plant field guides in an effort to tell an aster from a black-eyed Susan and a box elder from a black gum. Early on, I did learn to recognize the sassafras trees that were gathered at the wood's edge, because of the

Finding the Way

assorted shapes of their leaves, as if the tree has been dealt a mixed hand of clubs, diamonds ... and mittens.

I'm wearing a compass around my neck, but I'm not sure I know how to use it. I have a terrible sense of direction; I've been lost in my own woods, wandering blindly, trying to find the break in the old barbed-wire fence where I crossed over from the meadow. It's invariably only about thirty feet away—the opposite way from where I'm headed. My penciled notes tend to hedge: "north-ish" and "east/southeast??" B. kept calling a particular gate the "south gate," and I kept saying, "Isn't that west?" Ever the Eagle Scout, he tried valiantly to teach me how to read the compass, but I was sure it was west, because there were the mountains right beyond it. Blue Ridge equals west, right? But the Blue Ridge curves, like I'm seeing it through a fish-eye lens. Once I positioned the compass correctly, it was clear that the south gate *was* indeed south, and so was a natural spring.

I mark the spot where I first arrive at the meadow, at the top of the mile-long dirt and gravel road that leads up the mountain. There's a grassy farm lane that meanders from one end of the meadow to the other, up and down hills, following the crescent shape of the open land. Here and there, narrow paths shoot off the main one; some of these were first blazed by deer or bears. Because it would be a long time before we built a house here, we'd pitched a tent near the main path for shelter and shade under what turned out to be black locust trees, which have small thorns, and a honey locust, which has crowded clusters of scary spines, as long and thick as a pencil, that have been known to puncture tractor tires. Now I recognize the locust leaves, too: short, blunted ovals.

I note what remains of the cherry trees from an old orchard as either "healthy" (lots of leaves) or "dead" (none). I recognize these trees because they bear fruit, but also because they resemble their cousins, the cherry blossoms that have been planted in neighborhoods around Washington, D.C. There were once thousands of fruiting cherries here on the mountain; now there are only thirty or so. Some are too overgrown with vines for me to see whether any of the leaves are their own.

There are plants I can't identify, even with the help of the books; a tree I see along the edges of the meadow with branches that form droopy, palm-like umbrellas reminds me of the mimosa that grew in my childhood yard, but that's not what it is. The leaves are longer and larger and pointy on the ends. A large shrub with silvery leaves grows far off in the midst of the wild meadow where no path will take me, and I've been told to stay out of there when it's warm because of rattlesnakes, especially when all I've got on my feet are soft leather hiking boots. A giant tree with thick, twisty, rangy branches and large heart-shaped leaves stands alone in the field. It must be old to be as tall as it is, but what could it be? I make cryptic doodles on my map to mark the locations of these mystery plants.

I sketch the giant boulder that stands alone at the edge of a field near the gate to the south meadow. It was once known as Sentinel Rock because deer hunters used it as a lookout, but we're not hunters; not yet, maybe not ever. We call it Dominion Rock. Stand on top, and you can see for miles.

On a small hillside near the gate, I note "pink flowers." Many of the early blooming flowers here are weeds, but those pink flowers are wild geraniums. Before I came here, I only knew the geranium as a potted plant on my grandmother's windowsill. This was the first native meadow flower I identified using one of my pile of field guides. Achievement unlocked! I was pretty proud of this, even after one of my friends (a gardener, of course) said, "You didn't recognize a *geranium*??"

A plant with ferny leaves and tiny white flowers gathered in clusters that, from a distance, look like one big flower, turned out to be common yarrow. Yarrow, a book tells me, is a native flower in the aster family that blooms from late spring until fall. Not all varieties are native to North America, but the common nonnative one is so similar to the native, *Achillea millefolium*, it's difficult to tell them apart, and, apparently, most people no longer try: numerous sources describe the plant as "native to Europe and North America." A year later, I'd see it growing all along the mowed path. Around the locust trees, I find yarrow that's already two feet tall.

Finding the Way

My route continues past an offshoot trail that goes to a natural spring and stream and a shapely Virginia dogwood. A steep hill that leaves me breathless takes me to the cemetery and a large persimmon tree that was planted at its entrance by a previous owner. I learn to recognize the persimmon, when it's not fruiting, by its bark, which reminds me of charcoal briquettes. I walk on toward a majestic old white oak that stands in a clearing at the top of the highest hill at the south end of the meadow.

All of this rudimentary mapping has a practical purpose, to familiarize myself with locations of plants and landmarks on the mountain, with the lay of the land. But it has another, deeper purpose. While the less-than-quarter-acre I was born into—the willow, the mimosa, the silvergrass, the small hill where I could peer over the back fence—is forever imprinted on my mind, it will be years before I can claim an intuitive knowledge of the patterns and rhythms of a two-hundred-acre mountaintop. My external attempt to map this land is a way for me to try to speed up the process of internalizing it, to render it one of the few places ingrained in my memory when I think of home.

* * *

Wherever people go, they reshape the land, and on this mountain it's no different. Two hundred years ago, many of these hillsides were timbered for orchards. Around forty years ago, most of the orchards were replaced with cattle pasture. In the past decade, farmers stopped grazing cattle on the mountain. Haying eventually stopped, and mowing stopped, too. For a while now the land has been subject to its own rules; once people stopped managing it, it began to reshape itself. An old pasture filled in with blackberry. The forest encroached on the meadow, pioneer poplar, locust, and sassafras saplings taking the lead. A hayfield filled with hardy natives—and with weeds. Invasive shrubs climbed up and over a hillside, relentlessly expanding their range. Rather than a typical farm, the mountaintop looks unkempt, uninhabited, overgrown. The tendency might be to leave such a place alone—if humans stop interfering, nature will do what nature does, and take over. The land didn't need me; it would rewild itself—right? Not exactly. Once a place has been disturbed by humans, the only way to repair it is to keep disturbing it, but in the right way. The tricky part is figuring out what that means.

When I was nine, we moved to a new neighborhood, to a newer, larger house with a bigger yard, and my mother planted a vegetable garden on a sunny hillside. She buried a fish head under the tomato plants for fertilizer. The green beans and squash were attacked by slugs, so she placed a shallow pan of beer in the garden to attract and kill them. I helped weed and water and pick the vegetables, and while the garden was entirely my mother's doing, I was proud of the endeavor and the outcome. I was amazed that we actually grew vegetables we could eat. Sure, the corn was sometimes wormy, the green peppers were sometimes bitter, and the carrots could be short and stumpy, but they were ours.

Years later, I had my own yard in the suburbs, but it lacked a sunny spot for growing vegetables. Instead, I planted flowers and they were eaten. I planted flowers and sprayed them with garlic spray to repel the deer and the rabbits. It rained, and the flowers were eaten. I planted more flowers and stuck a coyote-urine-soaked popsicle stick I bought at the garden store into the ground. The stick lost its scent, and the flowers were eaten. Finally, I purchased a container of dried blood, the mark of the carnivore. (I couldn't help wondering whose blood it was. An unlucky bunny?) It came in a tall cylinder with holes in the top, like a container of finely grated parmesan. I sprinkled dried blood around the border of the flower bed like I was topping a pizza. The flowers were not eaten! Despite this small success, the supposed perennials didn't return the following year. (Fortunately, B. is skilled in gardening and landscaping; if left to me the yard would have been forever devoid of aesthetic interest.) I wasn't sure what I did wrong; I only knew that I was a failure at this gardening thing. And more important, I wasn't enjoying it, although I could see how one could easily become fixated on solving the problem of the devoured flowers, or the problem of grass not growing in shade, or the slugs that

Finding the Way

occasionally lacked interest in the beer that makes them shrivel up—and becoming fixated on problems like this was normally the definition of my personality. Instead, based on the advice of the late Polish philosopher Leszek Kolakowski, I assumed a theory of *not-gardening*. Kolakowski might have been lauded in his day for his three-volume history of Marxist philosophy, and, later, for his support of the Polish Solidarity movement, but I knew him for his satirical treatise offering nongardeners philosophical underpinnings to justify their choice, ranging from Marxism to psychoanalytic theory. My feelings at the time came closest to the Marxist option: you couldn't convince me that gardening for its own sake was a pleasure; that was a calculated lie promoted by the Gardening Industrial Complex. You couldn't fool me; gardening was work! I needed there to be a point to this labor beyond beauty—although I admit I enjoyed the beauty of other people's gardens. On the other hand, I could get solidly behind a vegetable garden, because then I'd be growing food. There was a functional goal. I know that people who garden take pleasure in the task of gardening for its own sake and for the results of their labor; I didn't. Of course, my results were not great. That probably didn't help.

I may not be selling myself as a natural for taking on a native plant restoration project, but only if you overlook my tendency to become obsessed with the nitty-gritty details that pave the way to a goal of questionable achievability. For me, this project *is* not-gardening. It's work for sure, but I don't see it as trying to cultivate or control nature so much as redirect it, as if the meadows are a bunch of toddlers about to melt down over a set of Play-Doh. My challenge is not to make myself necessary here, but to make myself useful. Someday the meadows will need to direct themselves; for now, it's up to me.

In school, I had a hard time learning exponents, because I'd hidden the fact that I didn't know the multiplication tables. I couldn't advance in math until I stopped counting on my fingers. Learning the mountain is a bit like learning exponents was for me. At first, I hardly knew what I was looking at; I didn't even know the right questions to ask. Despite my years of delving into everything from ants to tide pools to Andean condors, nothing quite prepared me to understand what was going on here. Even when I began to identify the species that live here and the relationships among them, I still had only a basic understanding, when what I needed was a next-level grasp of what was happening on this land. I could add, but I couldn't multiply. I needed help. I began talking with experts who I hoped could give me a crash course in everything I needed to know about the mountain: foresters and arborists, wildlife biologists and conservation biologists, bee and bird specialists, meadow restoration experts, hunters, livestock farmers, soil specialists and invasive plant specialists, master naturalists and all-arounders. I was sure these experts held all the secrets—the hard-science-based facts, the well-tested theories, and the predictions and insights that come from their deep knowledge of a subject and a place and long experience in the field. My education in native field ecology and restoration was about to begin.

Chapter 3

A NOT-GARDENER'S EDUCATION

... in which I'm introduced to a native prairie, a plant whisperer, and a no-holds-barred approach to meadow restoration

On an unusually hot, cloudless day in early April, I was standing under high-tension power lines on a dry, brown patch of what looked like last year's dead grasses and weeds, while a man talked intensely about native plants and geological phenomena and Sharpied diagrams on a whiteboard he held up with one hand for the group to see. Now and then, someone would call out a Latin name for a plant they thought they'd found. Others in the group would dart like minnows to gather around while a woman knelt on the ground and gently separated the dried brown grass with her fingers to reveal a green shoot an inch tall with maybe one leaf bud attached. The lecturer would lean over and take a look and affirm the find, and an excited buzz would spread among the group. This kept happening. I finally asked someone what the common name was for one of these plants. She told me that common names can be confusing because there's often more than one, so they always go by the scientific names. She was from the local native plant society. For a not-gardener like me, this was an intimidating scene.

This talk was presented in a nearby county by two local organizations focused on research and education related to native grasslands in Virginia. The topic was native grasslands and their history, and the seemingly inauspicious spot where we stood in a power-line right-of-way had been identified as one of the most diverse and exceedingly rare native grassland remnants in the state, likely more than one thousand years old. It wasn't planted; it had grown entirely on its own.

Any land that has been plowed in the past few hundred years—that is, any area that has been subjected to European farming methods—won't contain enough natives, or will contain too many nonnative plants, to be considered a primary grassland like that fortunate power-line patch. I'm only starting to learn about the importance of soil—how what's done to it can change what's in it, and how that can change what grows in a place, perhaps permanently. Plowing, which involves turning up the top several inches of the soil, ruins that soil in a bunch of ways. It destroys organisms living there (fungi and bacteria and insects) that are crucial to maintaining the nutrients that aid the growth of plants. Plowing exposes the deeper soil to sunlight, which kills those organisms and makes the soil more vulnerable to erosion by wind and water. Soil loosened and damaged by the plow eventually gets packed down and hardens unnaturally; then it can't absorb the water or nutrients it needs, and runoff runs rampant, washing away even more nutrients. Plowing was a big part of the destruction of midwestern prairies that led to the Dust Bowl, which sounds like a college football game played in a desert but was actually a devastating phenomenon in which repeated plowing, exacerbated by drought and wind, led to extreme loss of topsoil. The soil was so undernourished and dry that high winds were able to blow the dirt into dunes

A Not-Gardener's Education

tall enough to reach the roofs of people's houses. Depleted soil rendered the land a barren, empty place.

Before this talk, I would have thought a grassland was a grassland was a grassland. But the presenter defined any area containing some natives, but with significant nonnative or invasive species, as a secondary grassland. On a scale of zero to, say, seventy, with the higher number being the higher quality grassland, both horse pastures and Walmart parking lots will score near zero, because they contain few or no native plants, that is, almost no plants that originally grew in that place. (Gee, thanks for ruining horses for me.)

This rare native remnant of old prairie we were studying scored between fifty and seventy on the scale; it was a primary grassland. While secondary grasslands can be planted with native plants, they'll never become primary grasslands. For that you need to start with a true remnant of native grassland, where the soil organisms that support those specific plants will have evolved alongside them over thousands of years. Despite the native plants that persist there, the fields on the mountain can never be primary grasslands, no matter what I do to encourage them.

There are landowners who will plant a multiacre meadow where there used to be a pasture (as I originally planned on doing) and then spend every day tenaciously weeding and tending it as if it's a gigantic backyard garden. (I was not planning on doing that.) In the conservation community, there are those who refer to this approach somewhat dismissively as "pollinator gardening"—treating a meadow of large acreage the same way one would a smaller backyard garden. This reminds me of the phrase "a counterfeit natural landscape," which Michael Pollan used to describe Central Park in his book *Second Nature*. Such places may be beautiful, but they're not exactly natural; humans build and shape them. And how can a place be self-sustaining if you have to constantly weed it? But is what I aim to do that much different?

Restoration ecologists specialize in repairing ecosystems that have been damaged by humans. As I learned that day, a quasi-pollinator garden may be the best they can do when it comes to installing a meadow in a field that has been plowed anytime in the past, where most native plants that might once have existed have long since been replaced (intentionally or not) or overcome by nonnative grasses. Some ecologists have tried moving soil from existing native grasslands to a place they're restoring, in order to inoculate the soil there with the fungi the native plant seeds will need in order to grow. I would think the introduced natives could still face stiff competition from whatever weeds are already there, because those seeds will be in the soil, too. This isn't a concern in a primary grassland, since the seed bank—the seeds waiting in the soil—should contain all or mostly native plants. In fancy professional lingo, my once orchard, once cattle pasture, once hayfield, is an "old field," and its seed bank will contain a mix of all the plants left from those endeavors. But among them, hidden and not so hidden, is a promising force of native plant species. Can I tip the balance in their favor?

***

It was time to remedy my ignorance of plant names. I needed to prove to myself that I had it in me to learn to recognize a mountaintop of unfamiliar plants. I was aware that I was fighting a sort of mental block about plants, as if my brain didn't work right around them. I was determined to start seeing them as individuals. How could I know what needed saving or what needed eliminating if I couldn't identify the plants I found, if the meadow remained a blur of green? I understood that correcting my personal brand of plant blindness would only be possible if I could learn to see the unique traits of each plant the way I did with mollusks or beetles or dog breeds.

A woman named Charlotte Lorick, who was with the Smithsonian Conservation Biology Institute (SCBI), turned out to be the key to developing my awareness. I came to think of her as a plant guru. The SCBI arm known as Virginia Working Landscapes collects data about plants and birds on private lands in the state to help inform their conservation research and education initiatives. I asked if they'd be willing to come out

A Not-Gardener's Education

and take a look at the mountain. That's how I met Charlotte, an expert in an exhaustive list of areas including field botany, biodiversity conservation, ecological restoration, and plant community dynamics. Where the apps were uncertain, my intuition flawed, and the guidebooks inscrutable, Charlotte knows the answers. She has what I consider the uncanny ability to distinguish plants even in winter, when to me they're not much more than dried brown husks. I'd send her puzzling photos, and she'd reply quickly: "That's purple love grass," or "That's jimsonweed." Charlotte surveyed the mountain to identify the native plants there and made a few trips out to walk the fields with me. As we walked, she helped me match names with faces, plants like little bluestem and indiangrass, which are important native meadow grasses; and bee balm, aster, prairie rose, and ironweed, flowers I now understand are also *forbs*, any nonwoody flowering plant that ... isn't a grass. Most people I've met avoided diving into the overgrown field in summer, especially without snake-proof boots, but Charlotte didn't hesitate. She'd spot a promising stalk from the path and, in wellies that she admitted weren't exactly snake-proof, wade right in after it to get a closer look. One day in late summer, she emerged from the meadow covered shoulder to knee with the Velcro-like seeds of the crownbeard plant, a five-foot-tall native wildflower that proliferates in the fields. She didn't care. She'd found a native thistle.

As I mentioned, a grassland is dominated by grasses and nonwoody flowering plants—those forbs. A grassland (or meadow, as I think of it) can contain a few trees, usually fire-adapted trees—trees that evolved to tolerate or even benefit from the natural periodic burning that historically occurred in native grasslands. Then it's considered a *savanna*: a field punctuated by the occasional tree where a lion can hide and wait to prey on a zebra. I used to imagine a scene like that when I thought of a savanna, but savannas were once found all over the southern United States (I guess the city named after one should have clued me in), including on this mountaintop. Bert Harris, a scientist and director of the Clifton Institute, a research center that studies native grasslands, told me there would have been regular lightning-induced fires in these fields that maintained open savanna here for hundreds of years.

On the other hand, if half of your meadow contains short woody plants, like blackberry and dogbane, a relative of milkweed, you have a *shrubland*, which is a step in succession between a grassland and a young forest—succession being the gradual natural progression from one type of ecological community to another. Native grasslands and shrublands are favored by ground-nesting birds like the northern bobwhite, or bobwhite quail, a species of concern here in Virginia and elsewhere whose numbers are down nearly 80 percent since the 1970s, coinciding with the overall loss of native meadows. Meadows with a high concentration and variety of native plants, whether grassland, savanna, or shrubland, also feed native bees and other pollinators, which in turn feed native songbirds and their young.

I was slow to accept the fact that you can't vanquish every weed on many acres of land. There will be weeds in most meadows. Frankly, I'm still learning to live with the idea. Before Charlotte found that native thistle, all of the thistles I'd come across were unwanted weeds common in old cow pastures and hayfields. It would be nearly impossible for me to eliminate all of them without killing everything else, too. The goal is not to end up with a pristine field of native plants. The trick is to keep the undesirable stuff from passing a tipping point and taking over, but aiming for a perfect state of being for a meadow is the road to perpetual disappointment. And yet... (Perfectionism and persistence, it turns out, are annoying siblings.)

I'm aiming instead for 75 percent native plants, a respectable goal according to the ecological restoration folks, but I didn't settle on that figure until I'd learned more about plants and the restoration process. The current condition here may be closer to 50:50, which makes it feel like a tall order. There's hardly a consensus about these target percentages; I've heard everything from 50 to 100 percent discussed, depending on the condition of the land and on whom you ask.

What I needed was a large-scale restoration plan, but I didn't know that when I started calling contractors who say they specialize in establishing

A Not-Gardener's Education

native meadows. A gardening column in the *Washington Post* warned that such work requires "careful installation and maintenance" and is "laborious and expensive." I should have known that if it's discussed in a city paper's gardening column, it's most likely an approach that's best for smaller plots. And indeed, many of these small contractors are used to working on a few acres at most. This isn't the way to deal with seventy-five acres of old fields. But at the time I was still thinking small, the way I'd started out, telling myself I'd tackle an acre or two, and see how it goes. My brain was stuck in the suburban yards I'd always known, rather than fully seeing the broad landscape at hand. My scribbled map might fit inside my grade-school notebook, but it couldn't contain or convey the true scale of the land.

On the phone, almost every contractor told me the same thing, that the first step to growing a native meadow would be to spray the field with herbicide and kill all of the existing plants. Even though chemicals were historically de rigueur for maintaining a lawn, I'd stopped using them years ago in my suburban yard, at first persuading B. to try substituting corn gluten meal for a weed suppressant (yeah, that didn't work, but the rabbits loved it). I eventually decided it was better to have a yard full of clover and to try to ignore the other weeds, if what we aimed for was something green on the ground that our kids could play on. I didn't want to spray everything on the mountain with chemicals, either, and I couldn't understand why people seemed to automatically advocate their use in this wild place. Without even seeing the meadows, they told me it was the only way, that otherwise native plants could never compete with what was already here. Even after a visit, one man told me that spraying everything first was the only way he'd ever done it. Isabella Tree had initially sprayed a field twice before seeding when she first worked to establish a meadow. Her goal was to starve the soil, or defertilize it, eliminating the effects of years of overtreatment, since native plants are generally used to poorer soil than introduced grasses and crops.

The idea was, a few weeks after spraying, you'd have a big patch of ground with a bunch of dead plants. Then you'd seed this empty field with natives. The seeds would be planted using a method called *drill seeding*, which is said to be less disruptive to the soil than plowing. Many farmers have moved to such no-till or low-till methods for seeding and planting to prevent erosion and loss of valuable topsoil, and another Dust Bowl. In drill seeding, furrows are made in the soil by a machine, which then deposits seeds into the furrows and covers them over with soil.

A contractor suggested I check out a website for a mail-order seed company where I could see which seed mixes are available and how much they cost. The seeds themselves are, I was told, the most expensive part of the project. I was surprised to learn that there are potential problems even with planting seeds that come from native plants. Many of the seeds or plugs—seeds that have been started in potting soil—that are available for purchase are hybrids or cultivars. A cultivar is a plant that was deliberately developed by humans by crossing plants to select for specific traits; a hybrid results when two cultivars are intentionally cross-pollinated, again to produce an offspring plant with desirable traits. It's easy for me to think in terms of dog breeding. Standard poodles were bred to select for specific traits, which, early on, would have been the ability to retrieve a bird from the water. Those dogs were eventually crossed with the dogs with the most impressive coats and the ones that were most intelligent, to try to get to the best all-around dog. (They must have also selected for persistence, I think, as my dog once again carefully places her ball directly on my laptop keyboard.) Poodles are not wolves, and cultivars and hybrids are not the same as wild native plants.

Introduced cultivars can cross with wild plants, and that's almost always bad. Like the standard poodle, cultivars are genetically limited by definition; when a native plant is crossed with a cultivar and produces a hybrid, the genetic diversity of the wild plant is reduced and its traits are altered in unpredictable ways. This already happens routinely when seeds or pollen from garden cultivars travel to nearby wild areas and produce unintended crosses with wild plants. As with so many other changes, the effects reverberate. What happens to the creatures that depend on those plants? The resulting hybrids may not be insect friendly; their leaves and flowers can end up in shapes or colors that repel or won't

A Not-Gardener's Education

accommodate native pollinators. It's like replacing the chocolate chips in your favorite cookies with corn kernels. No one is going to eat that.

What are the consequences of introducing plants that don't come from the established microcommunity of plants on the mountain? Choosing native, open-pollinated plant seeds—that is, seeds from plants that were pollinated naturally rather than by humans—isn't necessarily a foolproof approach when the seeds come from a different state or region; those seeds may not match the native variety that grows here. They might not even bloom at the right time for local pollinators. What if the native plants that have hung on here for hundreds of years slowly disappear until no one remembers they were here at all?

The gold standard is to use seed that comes from a place near where that seed is to be planted, because it will be adapted to the local climate and the soil, but it's not always possible to find those seeds. The Clifton Institute is working to remedy that problem. They've begun collecting seeds from local native plants and providing them to local farmers who then grow seedlings that produce more seeds. The goal is to eventually make seeds available for native plant restoration projects around the state. Similarly, Charlotte suggested I collect seeds from native plants that already grow in the fields here and sow them in other parts of the meadow. That was the best idea yet. So ... I just need to figure out when the seeds are ready to collect, and whether I can scatter them or they need to be started indoors or they need to be cold stratified—placed in the ground or in my fridge through winter—in order to grow in spring. And, I need to make sure I'm taking seeds from the right plants. Every piece of information I gather leads to more questions.

Whatever seed is planted here, and by whatever method, in early spring when those seeds start to sprout, they'll face stiff competition from the sprouts of annual weeds whose seeds—a gazillion of them—have been waiting underground for their chance. Those seeds can survive in the soil for years, waiting for a disturbance that gives them room to grow. Killing all the other plants in the meadow could prove to be their favorite disturbance of all.

As those weeds emerge, it would be on me to pull them or spray them—again. And again. And again. Year after year, until the seed bank of annual weeds is exhausted, meaning they mostly stop growing back. Some weed seeds can remain viable in the soil for as long as seven years. Realistically, this weed search-and-destroy mission would never end, because new seeds are constantly being introduced by wildlife, wind, water, vehicles, and people. Multiply that monitoring by however many acres I choose to restore, and I could well be walking the fields all day, every day, for the rest of my natural life, crouching and peering at seedlings, trying to decide whether they're native grasses or noxious weeds. I fear that my crouch-and-peer days are numbered. And even with improved plant-identification skills, I might not do much better than I had at spotting the tiny seedlings those eagle-eyed native-plant aficionados had found under the power lines.

No worries! If you can't DIY it, you can hire the restoration contractors to monitor their own work—but not surprisingly that maintenance is a pricey proposition. The cost of the work, including the seeds, mechanical seeding, and monitoring, can easily add up to many thousands of dollars, another thing I didn't realize when I started down this road.

Maybe the nuke-to-restore approach makes sense for those places that fall near the bottom of the native-plant prevalence scale—the symbolic and actual Walmart parking lots. But on this mountain, where there are native plants worth saving, is it really necessary to burn it all to the ground with glyphosate?

If only I didn't know anything about what might be possible here, I could gaze at those meadows and believe they're fine as is. I'm sure the meadow installation contractors were advising me as best they could—they came highly recommended, and they were knowledgeable and experienced in their work. Many of them were trained horticulturists. But before I considered spraying these fields all over, or buying seeds—before I did anything else—I decided to seek advice from people who didn't stand to make any money as a result of my decisions.

A Not-Gardener's Education

My next call was to the Natural Resources Conservation Service, a department that falls under the U.S. Department of Agriculture and offers farmers free expert advice and, sometimes, funding assistance for conservation work on their land. I was ready for unbiased, unequivocal answers to my questions about how to restore these meadows to a healthier condition. I was sure there must be an optimal approach, The One Right Way, and I was determined to discover it.

Chapter 4

MAKING A PLAN

... in which I ask the government for help, meet a bee-hugger, and learn that "baseline condition" is in the eye of the beholder

The bumble bees fly intently but unsteadily, like drunks on a misbegotten mission. Their routes seem random and enigmatic, and I feel like a hapless air-traffic controller, watching as two of them collide; one drops away while the other wobbles off. A bee circles B., who's standing in the tall grass taking a photo. It passes him and then returns to hover at shoulder height, finally diving toward the ground and coming to rest on his pants leg above his shoe.

"Bee," I say, and point. He steps forward, and the bee takes off. It was evidently a step in the right direction, away from a ground nest we can't see. I take the landing as a reminder. While we're here, we're part of the meadow, too.

The buzzing presses in from all around, resonating like surround sound at the movies. The hum crescendos as I draw closer to a wild black

Making a Plan

cherry tree that's in full bloom. The tree is around twenty feet tall and at least as wide. I watch the movement of the bees. There's no still surface; the tree is alive with sound and motion. I turn away and I see more bees headed toward the tree, toward me. I shift out of their path. In the path of bees is not where I want to be.

I'm the intruder here. Like a character from the future visiting the past, wherever I step, whatever I do, something will change because of my presence. I don't think I've ever felt this so acutely. I want to change this place for the better—but how many people have made such self-consciously noble and grandiose statements and then gone on to make things worse? I only have access to today's information; what if, in six months or six years or sixteen, I find out I've gone about it all wrong, that I've somehow ruined what I'm trying to save?

The word restoration implies returning the land to an earlier, "superior" condition, or its *baseline condition,* a specific point in its ecological history. But which earlier condition, which point is that? There sure are plenty to choose from. Our idea of the baseline condition of a given place is usually how that place appears to us in our earliest memories, but in *Wilding,* Isabella Tree reminds us that the baseline condition of any land is ambiguous and always dates back farther than our memory of it. I don't remember the end of the last Ice Age, but I do remember what the woods near my house looked like when I was a child. What did they look like before I was born? Before my parents were born? Before the house where I lived, a few blocks away from those woods, was built? It's easy to understand why the land feels permanent to us, but of course it hasn't always been this way; in many cases it hasn't been this way for even fifty years, much less five hundred or five thousand. So what are we talking about when we talk about restoration? What conditions are we trying to recapture? What's reasonable to try for? I can't take the mountaintop back to the late Pleistocene. How far back can I take it—to what it might have been like before parts of it were timbered over three hundred years ago? Conditions here have changed dramatically over time; there were no invasive plants or insects then, and the soil organisms were different. How to account for those changes? Is moving the land backward in time even the right goal?

It's understandable that most people don't associate Virginia, or the South in general, with grasslands; most of us, when we think of prairies, think of the Midwest. But research supports anecdotal reports left by European travelers in the South who described wandering through seemingly boundless plains where the plants grew taller than a horse's shoulders. As early as 1540, Hernando de Soto reported wide open land in the South, not only dense forest. In 1670, German scholar and explorer John Lederer stumbled on meadow-filled valleys just east of the Blue Ridge, while searching for a route to the Indian Ocean (he may have been a little lost...). In his expedition report, he described the beauty of the savannas in early June: "Their verdure is wonderful pleasant to the eye, especially ... having travelled through the shade of the vast forest, come out of a melancholy darkness of a sudden, into a clear and open skie ... and flowry meads." And in the 1720s, the English naturalist Mark Catesby reported seeing "spacious tracts of meadow land ... with grass six feet high."

There's now compelling fossil evidence, too. Grass pollen has been underrepresented in pollen fossils because it's fragile, compared with pollen from woody plants, so for a while it was assumed that grasses never grew in such places. But microfossils, known as *phytoliths*, survive longer than pollen and bones, and they stay in place, rather than scattering like pollen. Through these phytoliths, scientists have identified plants down to the genus level, which, in the system of biological classification, is only one level above species—solid evidence for the prior existence of grasslands. Fossil soils provide more clues, helping to identify and date ecosystems, while fossilized animal teeth hint at what the animals that lived there used to eat.

Bison and elk used to roam the grasslands, maybe even here on this mountaintop. Fossil evidence has placed bison at high elevations, not only on low-lying plains. Bison ramble through fields, graze in mobs,

Making a Plan

trample the earth, roll around in the dirt creating wallows, and then they move on. Their wallows collect water, and the disturbed soil helps plants take hold. They disperse seeds, and their hooves break up soil and bury those seeds. In the 1700s, wild bison may have been more common in Virginia than anywhere else in the country, but by the late 1800s, they had disappeared from the state, along with the grasslands that were their habitat. The plants that had evolved to thrive where the buffalo roamed didn't evolve alongside cattle, whose grazing behavior is not the same.

Grasslands were once maintained by natural lightning-induced fires, long before fires were set by humans, and native grassland plants evolved to withstand and even thrive with burning. Human-set fires might imitate those natural fires, but they won't support the growth of plants that aren't already adapted to the existing soil organisms. That's why it's nearly impossible to create a self-sustaining native grassland where there never was one. Some southern prairies persisted without fire; they were inhospitable to trees due to their location, soil content, the volume of water, or soil that would freeze and slice through a seedling's roots before a tree could mature.

Although there may not have been as much open meadow here on the mountain before the early 1800s, when the land was cleared to plant orchard trees, the old oak that stands on the southeast peak in the meadow undoubtedly grew up in the open. This is clear from the way it has grown—wide, for stability against strong winds and exposure to harsh weather, rather than tall. If it had grown up surrounded by forest, it would have needed height, not girth, to reach the light. It may have been planted deliberately in a position of honor to overlook the valley and farm below. Or, it may have grown there by chance from an acorn buried by an animal. The tree may be three hundred years old, according to the arborist who examined it. That would mean the fields around it were already clear by the 1720s. Parts of this mountaintop likely existed as savanna for far longer, especially in places where the meadow is punctuated by wide and numerous rock outcroppings, and where more rock hides below a thin layer of soil, a fact I'm confronted with whenever I try to pound a stake into the ground.

To me, restoration will mean moving forward, not back, to strive for the best case of equilibrium for plants and wildlife, based on what, realistically, can grow well right now on the mountain, given everything that has happened to it in its history.

As humans, we tend to think of time in terms of our own life span, and the baseline condition of a place is often nothing more official than the conditions we remember. We return to a childhood haunt and lament how it has changed. The Chinese silvergrass that was my mother's pride is gone now, as are the mimosa and the willow. The tiny backyard where Rex ate the roses is now swallowed up by a swimming pool. I don't know what was there before the house was built and my once-favorite tree was planted, but the odds are against returning that land to its natural state, a state I never knew.

***

The Natural Resources Conservation Service (NRCS) was originally called the Soil Conservation Service, and it was started specifically in response to the Dust Bowl, to work to prevent soil erosion all over the country so that nothing like that would happen again. Although the organization was briefly part of the Department of the Interior, it almost immediately moved to the Department of Agriculture (USDA). Interior focuses its efforts on management and conservation of public lands, whereas the erosion of the Dust Bowl was heavily impacted by farming methods on private lands, so the move to Agriculture made sense. As part of the USDA, the NRCS supports conservation initiatives for farmers, through advice and funding. For instance, one of their programs compensates farmers for acreage set aside for conservation, separate from any land used for food production, and advises them on how to manage that conservation land effectively. The NRCS might also help to drill wells and install fencing on farmland to keep cattle from fouling streams. In most cases, some part

Making a Plan

of the land continues to be a productive farm, and NRCS program goals take that into account.

The mountaintop is a farm, and my hope is to someday use part of the land for food, whether that means raising chickens or shiitake mushrooms, if I can do it without harming the meadow I'm working so hard to save. I'm already growing more than 130 acres of timber, which I have no plans to harvest. But my first and most important goal is to restore the native meadows. Before I met with the NRCS soil specialists and biologists, I wondered how the fact that they fall under the umbrella of the agency whose central focus is agriculture might influence their perspectives. Would their advice be geared more toward protecting cropland than sustaining wildlife? Their orientation does come with some caveats, which gradually became evident to me, but my early meetings with them largely allayed my concerns.

One spring, Celia Vuocolo, then a private lands biologist with NRCS, came out to explore the mountain and offer her advice. Based on what she saw, she produced a detailed plan that I could follow to promote native meadows on the mountain. Celia specializes in habitat for the bobwhite, and she's an expert in native bees—an official bee hugger, according to the bumper sticker on her hatchback. She's cheerful, upbeat, and no-nonsense, despite the sometimes disappointing news she has to impart about the condition of the land. I can almost imagine her delivering the morning news, and in fact, on one of the days we met, she had just appeared on the *Today Show* to talk about the plight of a threatened bumble bee, one that she suspects may make its home on this mountain. Celia grew up in a rural township in New Jersey, and her affection for the countryside easily transferred to rural Virginia. She knows more about bees than I can claim to know about any one subject—and bees, as I come to learn, are not only one subject but many interconnected ones. As we walked the meadows, Celia paused to lean over flowers and peer at oblivious bumble bees.

I asked if she worried about getting stung, but she told me bumble bees aren't usually aggressive when they're on a flower, because they're drunk with nectar. And because they're part of small nests of fifty to a few hundred individuals, bumble bees are generally not aggressive and won't sting unless you directly disturb their nest. Honey bees are more likely to go on the offense because they have a much busier hive to protect: a single hive can house tens of thousands of bees. Most of the 460 species of native bees in Virginia have no hive at all; they're solitary, and because they have no nest to defend, solitary bees can afford to be more laid-back.

My brother was once stung by a bumble bee; he had picked it up off a flower because he wanted to hold it in his hand. He didn't understand why that was a bad idea, until he did. That was the start of his life as a pincushion; he was never stung by another bumble bee, but every manner of hornet, wasp, and honey bee seemed to view him as a threat. He became an "expert" on stinging insects, learning their names and talking about them incessantly. Perseverating, as it's called, is one of the effects of his disorder; he did this with anything he feared, often to the exclusion of any other subject. It's easy to see what my brother and I have in common: although I don't have Williams syndrome, a disorder caused by a random genetic mutation, I do have a tendency to obsess about a subject to the point of letting it take over my life.

Celia suspected there were two important bee species living on the mountain: the bee she discussed on television—the American bumble bee, aka *Bombus pensylvanicus*—and the black and gold bumble bee, or *Bombus auricomus*. In September 2021, the U.S. Fish and Wildlife Service found that there was reason to be concerned about the American bumble bee. An important pollinator for crops and native plants, it had disappeared entirely from eight states and had declined overall by 90 percent. The department announced it would perform a twelve-month assessment of the bee, after which it would make a recommendation about officially listing it as endangered. In early 2023, they announced that their "twelve-month" assessment findings would be completed ... in 2026. (I

Making a Plan

want to take five years to do something and pretend I did it in twelve months. I got so much done that year!)

The black and gold bumble bee, although listed as "least concern" at the moment, is headed in the wrong direction, too, having declined more than 50 percent since the 1960s. Celia warned me that it's not easy to figure out which of the two species she saw, and in a place this size, it could easily be both. The two are difficult to differentiate with the naked eye, even for a bee hugger; she'd have to examine their DNA to be sure. (This doesn't stop me from staring intently at the fuzzy bodies of random bees, as if I can parse the difference, if only I concentrate hard enough.) Until or unless I can get confirmation in a lab, I'm behaving as if both *Bombuses* make their home here on the mountain.

Contrary to what the meadow contractors had suggested, I learned that both drill seeding and another seeding approach known as discing would be a bad move on this dry, highly erodible mountaintop. Discing can lead to erosion by damaging valuable soil fungi. The central ridge of the mountain, where those two camel humps on the topo map meet, marks the division between two watersheds. The west side of the mountain drains into one river, and the east side drains into another; both rivers drain into the Rappahannock, which flows into the Chesapeake Bay. Standing near the old oak, I can see a glimmering stretch of one of these rivers in the valley below. Disturbing the earth on a large scale in order to plant seeds, even a few acres at a time, could lead to erosion of the rocky soil on these hills. The topsoil could wash away, carrying sediment and other pollutants into those waterways.

Lucky that, according to Celia, seeding wouldn't be necessary—there were more than enough native plants here, but they weren't "expressing" themselves. (Maybe what they really need is a blank journal with a whimsical cover, and a fountain pen?)

What that meant was, the plants were being stifled, literally. Years of dead plant matter had built up on the ground from normal seasonal die-offs of the profuse nonnative grasses and from mowing. This was all fine if I wanted a field of forage for cattle, or if I wanted to cut hay; the most abundant hay grasses here are smooth brome, timothy, and false oat-grass, all good for livestock, but no good for the native plants I wanted to encourage.

One of the most prevalent nonnative grasses here is orchardgrass, which is native to Europe and was brought over in the 1700s. It's probably been here since apple orchards were planted on the mountain. It grew well around apple trees and provided forage for livestock. Like most nonnative grasses here it grows in the cool season, but even then it appears earlier than most, its root system is deep and dense, and its growth outpaces most other plants. Others, like tall fescue, spread through rhizomes underground. If you were to dig up a section of sod in the meadow, you'd find a tightly packed network of roots holding the soil together. Most native plants have to first brute-force their way through that mat and the layers of thatch left by dead grasses in order to grow. Under these circumstances, the more delicate plants and those that emerge later in the season will be at a marked disadvantage.

When you gaze across the meadow, it's easy to pick out the nonnative grasses, because they crowd together like they would in a pasture or a lawn. Native grasses are predominantly bunchgrasses; unlike orchardgrass and other nonnative bunchgrasses, native grasses grow in clumps that sit farther apart, leaving open space around them for critters that need to navigate and find food. A meadow of native grasses can look airy and even sparse compared with a field of hay grasses. It's easy to assume that a dense, uniform green carpet of nonnative grass, the lawn or pasture look, is the healthier condition, but here on the mountain—and probably in most places in North America—that couldn't be more wrong.

I had good stuff here, as another NRCS wildlife biologist put it, but I had a lot of bad stuff, too, not only those hay grasses, but even more vigorous invasive plants that would take over if I didn't do something to stop them, and soon. I'd been advised earlier to mow one-third of the fields each winter, leaving two-thirds as habitat for overwintering wildlife, but

Making a Plan

now Celia cautioned me against mowing large areas of the fields in the absence of other crucial management steps; mowing would favor those nonnative grasses that come up earlier than native plants, it would make the thatch worse, and it could allow invasive plants more room to spread. For a while, "no mow" would become my mantra.

Celia proposed a restoration plan that detailed how to transform the meadow from an overgrown hayfield full of weeds and bramble, punctuated intermittently with native grasses and flowers, to a meadow that consists primarily of native plants—part grassland, part savanna, and part shrubland—where local wildlife could thrive. I was itching to get started (and that wasn't the poison ivy). The ingredients I needed were already here, dormant in the soil, waiting for a chance at some light and the space to grow.

The rough outline of Celia's plan went like this:

—Identify plants that already exist on the mountain.

—Identify, locate, and treat or remove the most aggressive introduced plants wherever possible to manage their populations and keep them from spreading.

—Slow succession by removing or treating saplings and other unwanted woody plants that creep into the meadow. Tulip poplars, a native tree, are the most flagrant offenders here. (There is nothing wrong with poplars in themselves, but here there are already too many, and their weedlike tendency excludes other, more ecosystem-important trees.) They're a pioneer species, which means they're the first to sprout up along the edges of the woods and make their way into the meadow, where they're looking to steal sunlight; they grow three feet every year, and if they're not stopped, there will soon be too much shade for sun-loving meadow plants to grow. That shade will in turn encourage the growth of more woodland plants and trees, until the meadow is well on its way to becoming a young forest. At the same time, the invasive emerald ash borer, an insect that kills the ash tree by feeding underneath its bark and cutting off its nutrient supply, has led to a mass die-off of the native ash in the woods here, opening up the canopy and making it easy for poplars and nonnative trees to plant themselves in those sunny gaps. More poplars in the woods means more poplar seeds planting themselves in the meadow, too.

—Promote soft or "feathered" edges, where relatively open meadow graduates to the denser shrubland that borders the forest. These adjoining habitats provide a natural transition between the sunny low growth of the meadow and the shade of the forest, offering shelter, food, and safety for birds and other wildlife that need one or two, or all three habitats to survive. Compare this with the typical yard, in which a cluster of trees meets a hard line of short mowed grass where very little can live. Instead, a gradual change in height, type, and density of vegetation provides a more natural wildlife corridor.

—After the frost, when native plants go dormant, mow or bush-hog only the fields that are being managed (bush-hogging can deal with taller, woodier growth better than a mower), and then treat with herbicide to kill nonnative grasses, which I should be able to differentiate because they stay green through the winter. I was assured that in the dormant season, native plants would be unharmed by spraying. The native plants in the fields include grasses known broadly as native warm-season grasses, whereas the hay grasses and fescue I want to minimize are nonnative cool-season grasses. The native grasses green up in late spring and early summer, and go dormant in late fall. Nonnative grasses will sprout up earlier in the spring, go dormant in summer, and revive as the weather cools. The fescue in my suburban yard (and on the mountain) is still green even in February, but in the dead of August, it turns thin, patchy, and sometimes brown.

I could divide the meadow into sections, and manage a few sections each year on a rotating schedule. Any fields I wasn't mowing or spraying in a given year would serve as winter habitat for wildlife.

But even as I heard the experts telling me that it would be tough to make progress here if I didn't spray any of the fields, I continued to feel uneasy about the potential for collateral damage, either by direct contact or by drift, in which the spray becomes an aerosol and travels on the wind. Some herbicides that are still commonly used were once repurposed for

Making a Plan

war, as defoliants and to ruin crops—like 2,4-D, which made up half of the dread Agent Orange. The nonprofit Natural Resources Defense Council has called 2,4-D "the most dangerous pesticide you've never heard of." How comforting. British ecologist Ken Thompson asserts that chemicals can cause more harm than the nonnative plants they aim to eliminate, by inadvertently killing native plants and disrupting other natural processes. But nearly every expert I spoke with, even those who generally steer clear of chemicals in their work, told me that because of the scale and combination of nonnative grasses and invasive plants growing here, I won't be able to restore the native meadows without them. I can be selective about which chemicals are used, and I can make sure they're used sparingly and carefully. (And I'll avoid 2,4-D.) I guess there's a happy medium between blithely ignoring the threat posed by noxious plants and flooding meadows with noxious chemicals, and I'll need to find it.

—In winter, burn the fields that were mowed and sprayed in order to eliminate the layers of dead vegetation, and to encourage the growth of native plants, many of which are fire adapted. Periodic intentional fires, known as prescribed burns, attempt to mimic the natural cycle of fire that historically promoted native plants.

—In spring, I'd sit back and watch native plants grow like magic in the fields that were burned. But on this point everyone agreed with the contractors I spoke with initially: I'd also need to be on a perpetual lookout for invasive plants.

—Do all of this again the following year ... and the year after that, and the year after that ...

—Once the meadows are healthier, I could add those sheep I keep daydreaming about. It's a moot point for now, since I don't have animals yet, but NRCS rules dictate that if I'm approved for assistance, I can't raise livestock on any acreage that's part of the program, even though some farmers graze their animals on native grasses. Even though some people bring in animals to intentionally overgraze nonnative grasses, to try to clear the way for native plants to grow. But NRCS is all about the tried-and-true methods, not the cutting edge or iffy or experimental, and there is something to be said for that. I was free to experiment on land that wasn't in a program, of course.

I gradually learned that nearly everyone around here with a small family farm applies for funding through myriad government programs available to those willing to establish and maintain native plants alongside their productive orchards or livestock practices. There would be a long wait to find out whether my application with NRCS was a success, and there were lots of projects competing for limited support. Even when a project is approved, the funding is often a cost-share model, so it only covers part of the cost of the work, and the participating farmer agrees to complete the tasks in the NRCS plan and make up the cost difference.

Patience is not my strong suit. I wasn't about to stand by helplessly for months while native plants in the meadow lost more and more ground to invasives. I decided that while I waited to hear, I'd start following the plan that was laid out for me as best I could. But there was only so much I could handle on my own. I set a budget and hoped it would be enough.

It's clear to me now why Isabella Tree was often waiting to hear about funding from one source or another before initiating various parts of her project. Not only was she working with thousands of acres of widely varied terrain and habitats, which she wisely divided up into parcels and approached in phases, but it takes a lot longer to do the work when you're on a budget. Before I truly understood what was involved, I thought that, outside of the obvious seasonal limitations, restoring a meadow was going to be a quicker, more streamlined, and much less expensive prospect. I thought I'd be able to work on one piece at a time, systematically, and still achieve my goals. What I'd expected to achieve in a year or two looks, more realistically, like a five-year plan, and even that may be ambitious.

Just as a farm needs a certain number of people working on it to keep it productive, restoring a large landscape requires people power. Knowledgeable advisers are crucial, but that advice isn't much use without people who can also *do*—who perform the actual physical labor of implementing it. Control freak that I am, even I can't do it all myself.

Making a Plan

My budget at the moment allows for a pretty limited number of hands, besides my own.

There's Brian Morse, a straight-talking wildlife biologist in an Indiana Jones hat who does ecological restoration, and who notably *did not*, at least at this stage, recommend spraying the fields and killing everything in order to get started. His team is willing to apply targeted, selective herbicide treatments on invasive trees and plants, in line with the conservative approach I have in mind. Then there's Adams, an industrious local farmer/excavator/property manager and my trusted adviser on all things farm-equipment oriented, who would eagerly bush-hog the whole place if I'd only give the okay. And, yes, there's B., the Eagle Scout, who will work for free, when I can drag him away from the new trails he's marking in the woods.

If the team is small, so are the tools. I'm limited to what will fit in the hatch of my seventeen-year-old Honda, the most prized of which is my weed wrench, the whimsically named Pullerbear; a long-coveted birthday gift, it even bears my initials welded into the stem. The weed wrench looks like a giant lever with jaws on the end. You fit the jaws around the base of a plant—anything from grass-blade thin to the four-inch trunk of a poplar sapling—and push down on the long lever, working the plant out of the ground as the jaws hold their grip. If the ground is wet, the weed wrench is said to make easy work of pulling problem plants. If the ground is dry, you risk breaking the stem and leaving the roots to sprout all over again. Here on these slopes, the ground is almost always dry. But one day after a rain, I managed to pry up eight problem shrubs, roots and all, and I felt like a Marvel comics superhero. A few days later, the ground was dry, and throwing my weight on the lever only left me suspended in midair, like when I was a kid and my dad sat down on the other end of the seesaw.

The weed wrench weighs twelve pounds and is awkward for me to carry, its jaws opening and closing and the two connecting arms levering at each other over my shoulder as I walk, which reminds me of hauling an unwilling toddler to the school-bus stop to meet his older brother. Sure would help to have a side-by-side, a small farm vehicle that looks a bit like a toughed-up four-wheel-drive golf cart with a mini pickup bed for hauling tools and brush. But that wasn't in the budget yet; first we had to fix the road.

When we first encountered the mile-long supposed-to-be-gravel road up the mountain to the meadow, for most of the way there were no ditches for water to drain off, and the water had found the easiest path, etching a gulley down the center of the road. The good part felt like driving through quicksand. The bad part, the final, steep, quarter-mile ascent, had long since washed away and consisted entirely of roots and rocks. The only way up the last stretch was on foot. Those who originally cut the road decades earlier had taken the most direct route, perhaps the only practical one, but after years of disuse, the mountain had taken the road back. Before any farm buildings, or the all-important farm house, could be constructed, the road needed to be remade into a proper gravel road with a crown and ditches and culverts to redirect the water, and check dams to slow it. It would be a year and a half before it was ready; until then, I did the best I could.

* * *

During economic downturns, when farms fell on hard times, they were often abandoned. Formerly tilled land, where plowing had altered the soil and eliminated grasslands in favor of crop fields, was abandoned to succeed to forest. The current inclination is to preserve such forests, as if they are the land's "authentic" condition—that inherently wiggly baseline—but in many of these places, this isn't the case. If no one is alive to remember what these forested lands were like before the trees, or before the farmed fields, it's as if the grasslands never existed.

At one point I wondered whether the logical plan on the mountain was to allow succession to ... succeed. That is, do nothing and let the meadow become a forest. I know that some of the fields were forest before the orchards were planted, but I don't know which fields or how many or

Making a Plan

where. And if I did let it all go to forest, it would mean giving up the entire sun-drenched meadow that has existed in some form for hundreds of years. It would lead to a big change in plant and wildlife populations: Ground-nesting birds that require shrubland and grassland would disappear. Thousands of light-seeking plants that attract native pollinators like those two special bumble bees would be gone, too, as would many of the other pollinators that now make a home here, like the monarch butterfly, which wouldn't find milkweed plants growing in the shade. The migratory birds that have returned every spring for, perhaps, centuries, might need to go elsewhere—but where, when native meadow ecosystems are in short supply all over?

Parts of the forest here were last clear-cut between the Great Depression and World War II. Since then, timber was taken here and there, as recently as twenty-five years ago, judging by the decay of the oak stumps in the woods. That means these are second- or in some cases third-growth forests. There's no way to retrieve the original old-growth forest once it's plowed under; the very definition of old growth means forest that was never clear-cut and has been left undisturbed. Like those primary grasslands, they weren't planted by anyone; the trees regenerated on their own for centuries. Old-growth forests support whole ecosystems of plants, wildlife, and soil organisms that aren't the same once the forest has been cut. They also store a lot more carbon than second- or third-growth forests. The younger woods here are filled with trees like those poplars that are less valuable for the ecosystem; they support a different and smaller subset of insects and birds, compared with the beneficial oaks they replaced.

But even Matt, a certified forester, didn't think I should let the meadow transform into forest. He assured me that the best way to protect the mountain's existing forests is to work on the meadow. The meadow, he told me, is what makes this place special. "Not going to find too many meadows on mountaintops that haven't grown up," he said.

Matt explored the forests here to get a sense of the place and what it needed. A bearded bourbon enthusiast and aspiring writer who enjoys sharing the wisdom he's gained in nearly thirty years of forestry, Matt is an enthusiastic translator of the messages of the woods. One summer day, we tramped through the woods around Heart Rock, our name for a fifteen-foot-high boulder formation resembling a heart and its chambers. I stepped gingerly over a low spot in the barbed-wire fence that used to keep cattle out. If only weeds would respect a barbed-wire fence the way cattle do. As we walked, Matt revealed the secrets to finding morels and the best places to plant ginseng. When I asked him if I should be on the lookout for copperheads in the leaf litter, he warned that saying "snake" when you're walking through snake territory is like saying "good luck" to actors before a performance: "You don't do it."

Near the forest's edge, he stopped to point out an invasive shrub and told us to get rid of it or it would multiply. "We did it to ourselves," Matt said, shaking his head over the general ignorance that led to some of the worst invasive plant problems that now face much of the world. The primary piece of advice he gave me was to control the plants that threaten the meadow to keep them from getting into the forest and taking over there, too.

I'm a writer, and I'm accustomed to using my imagination to predict what's most likely to happen in a story. I try to make what seems like the logical choice in a given scenario, considering both human behavior and external pressures. But the main characters in this story—weather and topography, plants and wildlife—don't follow the usual rules; add moderating factors: my personal limitations, the need to rely on others for help, and the influence of world events. (When I started this project, there was a pandemic going on whose impact no one could predict.) I can make all sorts of plans for this mountain, but I can't know how it will come out. I'm dealing with a living, breathing place and the interconnected creatures that make their homes here. The monarch butterfly, the soldier beetle, the dogbane, the bears, the locust saplings, the yellow-billed cuckoo, the deer, the old oak—they go about their business.

Making a Plan

Revitalizing a native ecosystem turns out to be more about stopping things from growing than I'd anticipated, a set of negatives required to achieve a positive. If I succeed, native plants will ultimately outnumber the weeds three to one; the land will support more native insects and provide food and shelter for grassland and shrubland birds. But when—if—that happens, it won't be time to rest; it will be time to start again. I get the feeling that no matter what, I'll always be running to keep up.

Chapter 5

ONCE UPON A TIME, FIVE HUNDRED MILLION YEARS AGO

... in which mountains rise and erode, apples come and go, and I question the meaning of land

When I told a local friend about the mountain, he sent me a video clip from the both famous and infamous 1939 film *Gone with the Wind*. When he was growing up in Virginia, he said, he lost count of the number of times he heard people repeat a line from the film—without irony—in which Scarlett O'Hara's father proclaims, "Land is the only thing worth working for, worth fighting for, worth dying for, because it's

Once Upon a Time, Five Hundred Million Years Ago

the only thing that lasts." Virginians, he told me, are obsessed with family legacies, especially when those legacies include land.

My family legacy, on the other hand, includes an alarmingly sharp carving knife brought over from somewhere in Russia, and a deep-seated fear of overnight hospital stays because my great-grandmother died in a freak accident in a hospital. But land? No. I never heard anyone in my family talk about aspiring to own land as a thing in itself, separate or different from a house, or a business. It still feels to me like a ridiculous and foreign concept, to own something like a mountaintop. It's not a pair of shoes, or a car. Where does such ownership begin and end? Do I own the soil and the rocks and the mosses? The toads by the pond, and the dung beetles, too? I can't quite get my mind around it.

I can trace my family back no farther than a pair of great-grandparents. The farm on this mountain comes with a much longer historical timeline, beginning before any member of my family ever hit America's shores. I'm only the third person to own this place since the Civil War. I can only know about those owners whose names appear on paper. There is evidence that Indigenous people came from the Ohio Valley thousands of years ago and established villages along the Rappahannock River miles from here, but I don't know who might have traveled from those riverside villages to hunt here, or whether they ever tried to make homes on this pile of rock. There are no archaelogical studies, oral histories, or anecdotal reports that I'm aware of that could tell me whether, or when, anyone might have lived on this mountain.

By the 1800s, this mountaintop was part of an estate that included thousands of acres in the valley. When I'm near the lone oak that stands on a peak above the far end of the meadow, I can see the old manor house that still sits a couple of miles away, and the cabins that housed the enslaved people who once labored there. The house is still owned by descendants of the same family that built it. By the time of the Civil War, the estate's holdings amounted to more than ten thousand acres of land, including the mountain.

BAD NATURALIST

From under that same oak, I can see the road where Robert E. Lee led his troops on their way to the war's deadliest battle, the Battle of Gettysburg, in late June of 1863. To the southwest, in the distance, I see the mountain where George A. Custer's troops fired on retreating Confederate soldiers who were returning from their loss in that same battle. Custer's troops were outnumbered by the retreating soldiers and were forced to escape to a nearby village. When those events occurred, the oak in whose shade I stand was already more than one hundred years old.

Beginning in the 1830s, it was time for apples. Mountains like this one, in the western Piedmont butting up against the Blue Ridge, were cleared and planted with orchard trees. These steep, rocky slopes weren't practical places for growing row crops like corn or wheat, but they were perfect for apples. The closer you get to the Blue Ridge, the better the growing conditions, and by some accounts, the Virginia Blue Ridge was a center of apple orchards in the early 1800s; apple trees were once planted all over the mountain slopes in what is now Shenandoah National Park. Here in the foothills, conditions are just right. Steep grades and rocky soil permit drainage, and good drainage prevents dampness, which in turn prevents naturally occurring fungi in the soil from attacking the fruit trees. Air circulation is better on slopes like this one, especially those between twelve hundred and twenty-three hundred feet. The prevailing wind blows through the orchards, helping to dry the foliage after a rain, and preventing dew from sitting too long on the plant. Cold air settles in lower elevations more densely, whereas slopes of moderate elevation like this one protect against frost. Even though it's colder at higher elevations, there's less frost because there's less moisture. European farmers cleared the hillsides on the mountain to plant orchards because they suspected that planting here would keep the trees from being killed by a late spring frost. One such event can destroy a whole season's crop, and not many farms, past or present, could endure more than one season of loss.

By the 1930s, I'm told there were as many as 150,000 apple trees on the mountain, stretching from the valley all the way up. Apple trees are crops; they can't replace the rich, varied ecosystems sustained by forests.

Once Upon a Time, Five Hundred Million Years Ago

A forest floor teeming with life can't give way to an orchard and still support that life. The fruit trees are not grown under a closed canopy, and they won't create one. They're planted in neat rows with open space between them and pruned regularly so they don't succumb to diseases brought on by damp, shady conditions. If dense vegetation grows at the base of the tree, it's sprayed to keep away moisture as well as critters that use the ground cover to hide from predators while they attack the tree. The trees themselves were once fertilized with manure from the orchardist's own cattle, raised side by side with the trees. These days apples are subject to more than sixty pests—many nonnative like the trees themselves—and require constant applications of pesticides and fertilizer in order to survive and bear fruit.

When I was in elementary school and we were stuck indoors on rainy days, teachers would dim the lights and show the class the same animated film, year after year, about the legend of Johnny Appleseed. A cartoon Johnny, a skinny wisp of a boy, hikes up bare hillsides with glee, carrying a stuffed-full messenger bag and tossing apple seeds like confetti as he goes. The seeds quickly sprout up, and empty green hillsides transform into thriving apple orchards before our eyes. But in reality, those hillsides weren't empty at all, and the real-life Johnny planted nurseries, not orchards; his goal was to sell trees, not fruit, and to establish his ownership of the land. He started planting in 1806, and he'd amassed an estate of at least twelve hundred acres across three states by the time he died. The story of Johnny Appleseed, a myth nevertheless based on the life of a real man, was eventually a public relations coup—apples were romanticized, and they became an American icon. American as apple pie. But, ironically, only crabapples like those Johnny planted are native to the United States. The apples from Johnny's trees would not have been eaten—they would have been pressed into hard cider to substitute for local water that wasn't safe to drink, same as many apples grown here in the early days. (Until methods to purify drinking water came into widespread use, people had a pretty good buzz on most of the time, courtesy of that cider.)

BAD NATURALIST

Tastier apples have their roots in the Tien Shan mountains in Kazakhstan, in central Asia. Their seeds were distributed on trade routes along the Silk Road, which eventually led to the hybrids we eat today.

There was a special variety of apple grown here in this county—the Albemarle pippin, a homely looking thing, mottled green and dull red, but it had an appealing sweet and tart flavor, and a great advantage: its flavor continued to develop after it was picked, and it was at its best in time for Christmas. Albemarle pippin trees once covered at least one of the slopes on this farm. Most of the apples grown in the county were exported overseas, and in the 1870s, a grower sent some of the pippins to the Queen of England. Queen Victoria liked the apple so much that to make it easier to maintain a reliable supply in England, she decided not to levy a tariff on her favorite apple. By 1930, the volume of most other farm products grown in the county had been surpassed by apples. There were more than three hundred thousand apple trees recorded here that year, and only around eight thousand head of cattle.

Fifty years ago, the apple business was in decline, and the second family to own the mountain decided to sell off some of their land. A woman from up north saw a tiny ad in a city newspaper offering the land at a low price. She bought the farmland in the valley and, a few years later, she bought the land on the mountain as well. Not many buyers were interested because it was difficult to access the mountaintop, and some local farmers predicted the land would never sell for more than $150 per acre.

The new owner built a house on the farmland in the valley. None of these previous owners made their home on the mountain. I want to, even as I can't help thinking that hundreds of years of people not trying to build so much as a run-in shed up here should give me pause.

Before European colonists arrived, a Siouan tribe, the Mannahoacs, is said to have lived in villages along the rivers in the Piedmont region of Virginia, but not enough is known about them. British explorer John Smith left an account from the early 1600s of an "interview" he

Once Upon a Time, Five Hundred Million Years Ago

conducted with a native person—whom he held hostage—a member of what he called the Mannahoac people. According to Smith's notes and observations, the Mannahoacs farmed the flood plains and lived just beyond them, along the river south and east of here. His captive is said to have indicated that the people avoided what would have been the Blue Ridge mountains to the west, because it was the territory of the Iroquois. But it's hard to trust that Smith's account is reliable. Smith's translator for his interrogation of the captive man was a member of an enemy nation and spoke a different language.

As much as I'd like to, I don't know the whole story. What does seem certain: the forests and savannas on the mountain most likely remained intact until Europeans arrived in the early seventeenth century and upset the balance, taking over land, clearing trees to use for buildings and heat, and planting crops by the same methods they'd used back home, without regard for differences in soil, landscape, or plant life that might have suggested a different approach, and without regard for the people who already lived in what became known as Virginia.

There are fewer than ten thousand people living in the county today. That number has stayed fairly consistent for the past hundred years. In 1850, the year the county was most populous according to available data, there were around nine thousand residents; nearly half of them were enslaved.

Beginning in the 1940s, the population would swell seasonally during the apple harvest, bolstered by the influx of migrant workers from Jamaica who were initially brought in to replace men who had gone to war. Those seasonal workers picked and packed apples on the mountain, and pressed the cider. As orchards here began to lose out to competition from those in Washington State, New York, and elsewhere, many of the county's apple orchard acres were converted to pasture. By 2017, the Census of Agriculture, which is tracked every five years, counted only 211 acres of apple orchards in the county, while more than 20,000 acres were in hay. Today, about half of the farmed acres here are dedicated to livestock pastures, mostly cattle.

BAD NATURALIST

There are nearly twice as many cows here as people, and products associated with cattle are worth three times as much as orchard fruits. This shift is reflected in the history of farming on this mountain.

After the apple business declined, part of the open land here was leased to a farmer who planted thousands of cherry trees. Rather than Johnny Appleseed, he was known as Cherry Gary. He installed an extensive irrigation system that reached far-flung parts of the meadow, remnants of which I continue to stumble on, and, for a millisecond, mistake for snakes. Cherry Gary abruptly abandoned his orchard on the mountain; whether because of a disappointing harvest or pressure from wildlife, I don't know. A few years later, he died on the single four-lane highway that runs through the county, when he lost control of his car and plunged into a gully. Around thirty of the cherry trees he planted remain; they've already outlasted their expected life span. When they bear fruit it feels like a sweet surprise—cherries I can pick from the trees and eat out of hand.

When the bears get to the cherries first, branches are strewn all around the grove like the aftermath of a storm. Broken limbs dangle from treetops where bears climbed up high to get to the fruit. The scat evidence is everywhere, too, full of cherry pits. But last year, there was oddly little damage, and we ate all the cherries we could reach. I suspect it's because the black bears are suffering; their numbers are down, and the Virginia Department of Wildlife Resources has announced there's an outbreak of mange.

The woman who owned the land before me raised cattle on her fields in the valley, and she was set on running a cattle operation up here, too. She bulldozed the apple trees on the mountain, leaving only one, which she saved as a nod to history. It stands alone in the middle of a field of orchardgrass.

* * *

The Blue Ridge mountains get their name from the way the light reacts to hydrocarbons that are released by the trees that cover the slopes,

Once Upon a Time, Five Hundred Million Years Ago

lending the mountains a blue cast. I see it now, from the top of this mountain, a watercolor palette of blues stretched across the horizon. This mountain I'm standing on was once part of the oldest mountain range in the world. Around a billion years ago—a time frame that's beyond my ability to conceive—a supercontinent gathered itself, tectonic plates colliding, and a long period of mountain formation ensued. Sections of the earth were shoved up and folded back on themselves, reshaping and reforming and then destroying each other again and again in order to create. I search for and do not find a metaphor, because mountain formation is too monumental; it *is* the metaphor. I imagine the continents rising, splitting, crashing, the land stretched and kneaded until, over hundreds of millions of years punctuated by periodic violent mountain-forming events, the land was utterly changed, and the mountain chain we know as the Blue Ridge came to be.

The tallest peak in the Blue Ridge is over six thousand feet, which seems diminutive compared with the Himalaya—Mount Everest is twenty-nine thousand feet high. But the Blue Ridge mountains were once that tall, too. Age wears mountains down. (I can relate.) The Himalayas are relatively young mountains at a mere twenty million years old; they'll eventually weather and shrink like the Blue Ridge, though I won't be around to see it. By two hundred million years ago, the Jurassic period, the Blue Ridge had eroded down to near its current size.

The upheaval that shaped the Blue Ridge continued for hundreds of millions of years—the fires, the rents in the earth, the storms, the volcanic eruptions—slowly transforming the aspect of the earth's surface. What would it be like to live through it? Although I know it happened too slowly to notice, I imagine daily drama, like the "Night on Bald Mountain" sequence in the film *Fantasia*, with its moody Mussorgsky score: the earth splits apart, mountains rise, a giant horned devil appears. When I was five, this was the scariest thing I'd ever seen, a formative experience for me, if not for actual mountains.

Geologists know that the Blue Ridge and the Piedmont date back one billion years, because Precambrian rocks are visible on the surface of the mountains. These rocks were once buried underneath miles of softer sediment that eroded away from the peaks over time, during subsequent mountain-forming extravaganzas called *orogenies*. (If you quickly read that word as "orgies" you're not wrong; you could think of them as orgies of mountain-range building.) In Virginia, the west side of the Piedmont bumps up against the eastern edge of the Blue Ridge, and the east side of the Piedmont stops at what's known as the Coastal Plain, bordering the Chesapeake Bay and the Atlantic Ocean. The entire Piedmont is far longer, stretching along the Appalachians from New York to the middle of Alabama. The word *piedmont* means "foothill," and that is where this mountain sits, at the foot of the Blue Ridge. The farther west you go in the Piedmont, the closer to the Blue Ridge you'll be, and the more small mountains like this one you'll encounter. The farther east you go, the flatter the terrain. But here on the west end of the county, even the valleys don't feel flat; the roads curve and wind, climb and plunge, and the mountain range seems much closer than a few miles away.

Some people call this mountain a *monadnock*—the name comes from Mount Monadnock in New Hampshire. A monadnock is a relatively small, isolated mountain that has resisted the worst erosion and stands alone. The Piedmont region of Virginia contains a number of these lone mountains amid the rolling hills and pastures as you approach the eastern edge of the Blue Ridge. Even though it's dwarfed by the Blue Ridge peaks, because of its position in relation to the mountain chain, the luck of geology, this mountain feels taller than it is. At nearly fourteen hundred feet, it is one of the taller "miniature mountains" here. It too has eroded; the exposed rock at the highest peak and on the forested slopes tells me as much. When I tested the soil composition here, I was supposed to dig down to a depth of eight inches to take samples. I crept around the slope digging small, shallow depressions, trying to find a place where my spade didn't hit solid rock only four inches down.

Visitors here have their favorite views. Most want to stand under the old oak tree and gaze south across the valley. They name the hills in the distance—other monadnocks, and at least one extinct volcano—and

Once Upon a Time, Five Hundred Million Years Ago

the farms below. Or, they stand at the northwest end of the meadow and look southwest toward the silhouetted Blue Ridge chain.

If you try to hike up some of the wooded hillsides here, you'll need to watch your footing as you pick your way over scattered rocks and broken boulders. Some of these slopes could be old boulderfields, basically fields of rocks, formed by long-term rock weathering combined with the melting and withdrawal of glacial ice. On those hillsides I find sweet birch, red oak, chestnut oak, and black gum, all species that might grow in a boulderfield. You're more likely to encounter these ankle-busting fields on the higher slopes of the Blue Ridge, in places where there may be few trees or none at all; low-elevation boulderfields, anything below three thousand feet, are relatively uncommon.

About a half-billion years after the Precambrian rocks that date the Blue Ridge were formed, there was a shallow ocean to the east of what is now North America. The plate to the east of that ocean began moving west, and the two plates gradually closed around the ocean like a vise. A long stretch of volcanic islands—a volcanic arc—rose up in the space between the two plates and was squeezed from the east toward a collision with North America. I imagine a catastrophic series of formative and destructive volcanic eruptions along the boundaries of the plates. These volcanic islands left their mark on the Piedmont slopes. The slate I occasionally find amid abundant granite here is a remnant of that collision, sedimentary rock thrust up from the bottom of an ancient ocean and metamorphosed by the weight of time.

∗∗∗

I'm afraid of heights, which seems like an odd fear for someone who plans to live on a mountain, even a small one. I've forced myself to move past this fear in manmade places, albeit in ways that may seem tame, like climbing the 551 steps up to the dome of Saint Peter's Basilica and looking out over Vatican City, while traveling alone in Rome; or peering out from the final stage of the Eiffel Tower, over nine hundred feet up, while on a student exchange trip. But high places in nature unnerve me. I've never seen Mount Everest except in films, much less attempted to climb anything high enough to require supplemental oxygen, although I have visited Cotopaxi, a nineteen-thousand-foot dormant volcano in Ecuador, when I stopped on the mainland en route to the Galápagos. A friend insisted on driving us up the mountain until we reached an altitude where the car stalled out for lack of oxygen. I emerged from the car into a snowbank. Whipped by an icy wind, I couldn't force the door shut, so I hung onto the handle, certain I would be torn from the spot along with the door and flung down the mountain. My brain was swaying in my skull, but my dizziness came more from the mistake of looking down than from the change in altitude. I didn't make it to the top.

On this mountain, I'm not worried about being blown away by anything except the views. Even so, the winds are strong, and the weather tends to extremes. A sunburn is possible well into November. Winter comes, and one moment it's bright, the next I'm in a blizzard of snow and hail; ice could be pelting my cheek, and a minute later that's done and it's sunny again, but with a cutting wind to remind me how cold I would be if the sky were to cloud over. I've never much liked the cold, although not long ago, when I could still run, I was known to do so before dawn in below-freezing temperatures. Wear a balaclava and slip hand warmers inside my gloves; keep moving, and, I tell myself, I'll be fine. When persistence teeters at the edge of obsessiveness. But that's another reason I might be able to make these restoration plans work. The persistence/obsessive line is where I live. How else can you stick with something for years with no promise that anything will come of it? That's what it's like to be a writer. And, I guess, to restore a meadow.

My family never had a reason to want things to remain the same. Going back as far as I can on my father's side, my grandparents lived in a village in what was sometimes Russia, sometimes Poland, and now Ukraine. My grandfather rode a horse in the army, and he proposed to my grandmother from atop a white stallion. They fled the village in

Once Upon a Time, Five Hundred Million Years Ago

1912 to escape the murderous rampages known as pogroms, and the certainty that my grandfather would be conscripted by the Russian army to serve in the horse-drawn artillery in the coming war, and that he would be killed. (Russian soldiers regularly made Jewish soldiers the targets of harassment and violence.) Any family of mine who survived the pogroms were later killed by Nazis who wiped out all the Jews in the village; only one of my grandmother's ten siblings escaped.

When my grandparents came to the United States, they opened a small grocery in Anacostia, in a majority Black neighborhood in the southeast quadrant of Washington, D.C. Around the time the apple business reached its peak in Virginia, my father was a child living behind his parents' store. Shopkeepers with more money lived in houses or apartments above their stores, not behind them. In the store, my grandparents sold mackerel, which they kept swimming in a tub of salt water. They sold five-pound bags of hard and soft coal and bundles of kindling; kerosene lamps and, separately, the wicks, and kerosene by the half gallon, which they transferred into containers their customers supplied. They sold loose cigarettes with which you'd receive one match; plus, chicken feed, bulk flour, and chitterlings; pig's feet and penny candy; and don't forget hosiery, for men and women. The store also sold apples, which my grandfather bought at the grocers' market in southwest D.C., along with most of the food and produce he carried. The apples undoubtedly came from farms in Maryland or Virginia. I don't know if he stocked those famous pippins; it was a poor neighborhood, and he would have stocked the least expensive apples available. My grandfather's family had owned a tavern in their village in Poland, but in Washington he tried perpetually to perfect wine made from black grapes and sugar, with no luck. According to my father, it was a vile concoction.

All of my grandparents and great-grandparents started and ran their own businesses. My parents eventually did as well. My father spent decades selling Remington typewriters, until it seemed like he was known to every procurement officer in half the government agencies in Washington. He worked on commission, and he'd come home with stories about the salesmen who drank with the boss at the local bar at lunchtime getting the best client lists. Having a father who worked in commission sales always gave plays like *Death of a Salesman* and *Glengarry Glen Ross* an extra dose of relatability for me, even though my father was never downbeat, or beaten down, like the protagonists of those stories; it's not in his nature. Around the time the apple trees on the mountain were cleared to make room for cattle, he went into business for himself. My family valued self-reliance and independence, and they worked hard to get it and keep it. We have this in common with people who farm, even if we never owned land.

My mother's was the more affluent side of the family, until divorce happened at a time when divorce was unusual, leaving my grandmother high and dry. My mother's family owned a beauty school. My uncle (my mother's brother) styled Lady Bird and Lynda Bird Johnson's hair. I spent part of my childhood wrinkling my nose at the smell of Aqua Net and eavesdropping on women who were sitting under hot dryers, gossiping in curlers.

I was five years old when the woman who'd eventually sell me this mountaintop first bought the farmland in the valley. Neither side of my family passed down land. My family legacy includes two old banjos and a tendency to worry.

In the newer suburban Maryland neighborhood where I spent most of my childhood, there's a tiny farm-family plot, a graveyard that dates back to the mid-nineteenth century, well before my family arrived in this country, before anyone thought of building that neighborhood on hundreds of acres of farmland in the first place. The handful of graves are preserved in a grassy liminal zone between the backyards of two houses that were built, along with the rest, in the early 1970s. I walked my dog in this small plot in daylight, trying to read the faded stones. The engraved words and the edges of the headstones were degraded and distorted by time and weather. It looked to me as if the stone had begun to melt, like it was made of wax. When I was older, my friends and I would gather in this graveyard on Halloween nights and try to scare ourselves. Eventually

Once Upon a Time, Five Hundred Million Years Ago

the people who lived in the houses around the graves built tall fences so they wouldn't have to watch their children swinging on a playset against a backdrop of tombstones. That family plot is the only evidence that remains of the farm, since the last generation could no longer afford to resist selling the land. A Levitt development of over five hundred houses sprang up where the farm once existed, where those dead, boxed between a one-story rancher and a Dutch colonial, had once toiled.

Everyone knows the mountain farm has new owners. "I heard someone bought the place," said an excavator, sizing me up. Everyone has been welcoming, but coming from a suburb that seems at times like a small city, it was a little unnerving at first to realize that people we didn't know anything about already knew something about us. I know we're being scrutinized; it's only natural. One neighbor expressed relief on learning about my conservation goals for the place she's always thought of as "one tree hill."

When our house is finally finished, I'll be living on a corner of this mountaintop, in a spot where I can watch the light in the meadow shift every day, as I imagined I would when I first saw it. That's a big change for this place, and for me. Even if we all stand very still and hold our collective breath, there's no stopping change. I hope it's reassuring that I plan to tread lightly here. I've never thought in terms of legacies, but maybe this is my version of one.

***

With most of the orchards long gone, I wasn't sure I'd see any apples on the one tree the last owner left for posterity. For apples to grow, the tree must be pollinated by bees that have collected pollen from a different variety of apple tree. I'm not sure if the previous owner knew this, or if she would have cared. She was farming cattle, not apples. But then one spring, perfect pink and white apple blossoms appeared on the tree, like a mirage, and in the fall, there was fruit. The apples weren't much to look at—small and misshapen, red with a pool of green around the stem. But they were tasty—the right balance of sweet and tart, like the ones I'd heard the Queen favored. How was the tree pollinated? One of the few remaining apple orchards in the county sits on adjoining land, hundreds of feet down the mountain.

Last year, the apple tree was attacked by tent caterpillars, their dense webs wrapped with geometric logic around branches and over twigs, reminding me of mangled kites. One tent became three, and three became five. I leave these nests alone when I see them on wild trees, but on this one decrepit apple tree, the last of its kind on the mountain, I'm afraid they might kill it. If I could, I'd aim the garden hose at the tents, but I have no hose hookup and, still, no water source. Instead, I take the longest rake I can find and wade into the field wearing tall rubber boots that offer no protection against a startled snake. I scrape at the webs with the rake's stiff metal teeth, splitting them open and scattering the caterpillars to be eaten by the yellow- and black-billed cuckoos that consider them a treat. The toothed end of the rake is heavy and hooks onto the branches, but I untangle it carefully each time without damaging the tree. I won't let this last tree be destroyed.

I lost track of time. I don't know how long I stood out there in the sun and, eventually, in the rain, waving the rake over my head. I didn't notice the soreness in my shoulders until I was done and had walked back down the hill to the shade of the tent and the locust grove and a jug of water and realized my arms were shaking.

From the high point of the cemetery across the meadow's undulating hills, the nearest peak of the Blue Ridge looms in the northwest, shadowy and lush, miles away but feeling as close as my hand. A cloud sits low, reshaping the mountain's peak into the mouth of a volcano. A turkey vulture passes overhead and then another and a third. I think of that string of volcanic islands that pushed up from beneath a shallow ocean that no longer exists, and everything that needed to happen over hundreds of millions of years in order for this place to become what it is, for

Once Upon a Time, Five Hundred Million Years Ago

these mountains to be mountains and for the creatures that live here to live here and for me to stand here watching as if I might witness the slow, imperceptible geological shifts that will continue as long as the world exists, when I'm the one weathering and eroding at a speed that doesn't require time-lapse photography to capture.

I want to leave something more lasting than a deed filed in the county clerk's office showing I once owned the land on top of this mountain. I know whatever I'm able to do here is only the beginning. The land keeps changing and the earth keeps spinning, and whatever lives here will be subject to the motion of continents and the insistence of weather and the resistance of rock and not the whims of people. Family legacies are a freckle on the face of time. No matter how much land we accumulate, and no matter how long a family holds it, in the end it isn't ours to keep.

PART II: THE MOUNTAIN OF WEEDS

Chapter 6

TWO WAYS OF LOOKING AT A HILLSIDE

... in which a farmer's weed is my wildflower

On an early visit to the mountain, I walked along a freshly mowed path with Alan, who is a sheep farmer and a real estate agent. He was the one who first brought us here to see what he called the best view in the county. We had no thought of buying the place before then, but that was about to change. (We needed a house, and that key factor was missing from the equation. But once we saw the mountain, our plans went out the unfortunately figurative window.) It was the dead of summer under a cloudless blue sky, and the heat pressed me down to earth, as if being a thousand feet closer to the sun made gravity more powerful, too. Alan wore long sleeves with the cuffs hanging loose, and long pants and

Two Ways of Looking at a Hillside

boots, an attire I've now adopted myself, after a lifetime of summers in shorts and open-toed sandals. (I'm still surprised to find that I feel cooler in long pants.)

The place hadn't been mowed in some time, until now, and this path had been cut solely to enable a visitor to walk from one end of the meadow to the other without fighting brambles and collecting ticks. Five years earlier, this was a cow pasture. Now it was so thick with growth, I could hardly see anything beyond the green wall. Bumble bees were everywhere, attending to a profusion of raggedy-petaled yellow flowers that towered over me and closed in from either side. It felt like humans didn't belong here. But I was drawn to the place, and I wanted to know it. I asked Alan what these tall plants that dominated the fields were called. He grabbed hold of one and snapped it off midstem, flicked it back at the meadow with a sigh.

"Stickweed," he said. "The bane of every farmer."

Alan has the mellifluous voice of a stage actor, and when he said "stickweed," his mild disgust sounded dramatic. "Cows won't eat it," he told me. Even goats won't eat it. When it dies, it leaves tough, square stalks sticking up all over the field.

Hence stickweed. I quietly freaked out—this terrible plant was everywhere!

"Bush-hog it," he went on, as if sensing my angst. "Spray it. Keep it mowed. It's not a problem."

Is it a problem or isn't it? Back home, I did my research, and discovered that the so-called stickweed can easily take over a pasture. I started asking around to find out what to do about this awful weed. I was told by agriculture specialists and gardening organizations that it was an aggressive plant, nearly impossible to eliminate without using chemicals, and it was a perennial, so treatments would need to be repeated annually until it stopped coming back.

At first I felt discouraged, but soon I thought to wonder, why *was* this stuff so good at growing on the mountain? I'd assumed it was a weed.

What if I was wrong? I'd been asking how to get rid of it before I knew whether I should get rid of it.

Everyone loves sunflowers, right? Years ago, I drove over winding roads between hilltop villages in southern Spain, past iconic fields of sunflowers, their wide faces turned toward the sky like a crowd of sunbathers. And who doesn't get cheered at the sight of a daisy? Do kids still ask "Loves me? Loves me not?" while sadistically plucking the simple white petals, leaving only the single one clinging to the flower's yellow button center, the answer?

There is no such embedded cultural attraction to the poor, maligned yellow crownbeard—aka stickweed—even though it's cousin to both of these flowers. The crownbeard, or *Verbesina occidentalis*, is a member of the aster family, but it's that relative everyone sees as a loose cannon—shows up late for dinner, belches loudly, and swipes food off your plate. Hard to like, but family is family.

The yellow crownbeard has all the wrong features of the more popular plants that share its genus: it's tall and yellow, sure, but it's no bright crowd-pleaser like the sunflower. Nor is it anything like its small, delicate cousins the asters.

Its sturdy stem looks square—or what the experts call *winged* for the flat flaps that run the length of its stem—and it can grow as tall as fourteen feet. In the meadow here, it seems to top out between five and eight feet. But there is little to recommend the aesthetics of the flower that sits at the top of that towering stalk. Unlike the daisy, its sparse, unevenly arranged petals appear perpetually half-plucked, like a lovelorn teen gave up in the middle and decided the question wasn't worth answering. That's how the flower grows, a bald, wispy thing.

I'm making it sound forlorn. But native bees and butterflies, like the silvery checkerspot and the gold moth, which lay their eggs on the plant, don't argue aesthetics. The soldier beetle hangs around the crownbeard, too, controlling insect pests by eating aphids, mites, and grasshopper eggs, while it helps pollinate the plant. Yellow crownbeard blooms in late

Two Ways of Looking at a Hillside

summer, when so many other flowers brown and shrivel—August and September, and even into October here. Late-flowering plants like this are crucial for late-season pollinators that are storing up energy for the long winter.

A stiff wind might cause the plants to lean, but they won't bend over or break without direct effort. The stalk is dense and strong and stays that way even after it's dead. You have to crack it like a stick the way Alan did if you want it out of your way. But I leave it for the insects that overwinter in its stem, and the birds, like the bobwhite, that feed on the seeds.

One reason crownbeard is so profuse here in the first place may be that this was once a pasture. If cattle are allowed to graze too long on one field, they'll overeat what they like best—here that means hay grasses, fescue, and orchardgrass—until the grasses are so low that, instead of resprouting robustly, the plants gradually weaken and leave less decaying plant matter behind. That in turn means fewer nutrients are returned to the soil. This lower-nutrient soil is what the native plants here prefer—it's the soil's natural state. Crownbeard's young shoots are edible, but once it grows up, cows dislike the tough stems, and they leave it alone. Take that low-nutrient soil and add a perennial native like crownbeard that isn't interesting to cattle, and the plant will seize the opportunity to multiply.

What plant is forlorn that can dominate a meadow the way crownbeard dominates some of the fields on this mountain? It's an energetic spreader, a weed for those who don't want it, but for those of us who appreciate it, it's called "successful." It's a big part of this native meadow ecosystem; it can push its way through the dense thatch when so few natives can. I find it sprouting up in the middle of the farm lane faster than we can keep it mowed. Yes, Virginia, there can be too much of a good thing, but its zealous presence here may mean fewer of the plants I don't want in the meadow.

In a webinar called "Winning the War Against Weeds," the Smithsonian Gardens calls yellow crownbeard "unsightly and aggressive" while acknowledging that it's a native plant. Unsightly? I beg to differ. Sure, it may not color between the lines, it may be rangy, but I'm building a wild meadow, not the gardens at Versailles. If a healthy garden can only be one that's trim and neat and not wild and free, if a gardener is pulling native plants as weeds and replacing them with exotic ones, maybe that garden isn't as nice as its keeper would like to think.

A native wildflower, and there were acres of it. I was beginning to understand that, like people, the plants here are not only one thing.

*　*　*

The following year, a fall morning dense with fog. Climbing the road to the meadows, I can see only a few inches ahead of my low beams. At the top, I enter a cloud. The song sparrow and the white-throated sparrow call from opposite directions, distant, shrouded. As I approach the field, the fog begins to lift, and the first color I see is gold: broomsedge amid a dried autumn bouquet of wild bergamot, skirting the edges of a wide rock outcrop shaped like a boomerang.

The rock is silver gray, covered with ivory lichens, and punctuated by coyote scat. (Gray and hairy, the scat is easy to spot and reveals that Wile E. and friends gather here regularly.) I was initially fooled by the size of the outcropping, which winds down a gentle slope for around twenty feet, then ducks under an island of grasses and reemerges on the other side, and I realize that another flat outcrop I'd thought was separate is linked to it beneath the soil. Rock appears to plate this slope like an exoskeleton. If I excavated, I'd find a boulder of a size that Sisyphus might find daunting.

This time of year the air smells sweet as the crown of a newborn's head. The mist lifts slowly and the topography is gradually exposed. The meadow with its hills and valleys lies between the mountain's two high points, like the dip between the horn and the cantle of a saddle. I stand at one end, on the horn. The big old oak tree is at the far end, the tip-top of its corona just visible above the cantle. The distant north-facing hillside far across the meadow from me is the steepest slope within view. It's late November and the hillside is golden like the flank of a lion; I can make out the shape of a golden shoulder on the ridge. The light glints off it.

Two Ways of Looking at a Hillside

The hillside is a half mile from where I stand, and as I make my way toward it, I reach a low spot in the meadow, and the place I came from all but disappears. I'm in the bottom of a bowl, and the hills that rise and fall around me, the tall, dried remains of crownbeard and curls of bare bramble obscure my view. Hiking up the path to the ridge that marks the southern high point, I lean into the grade. I walk serpentine because it's easier than climbing straight up.

Near the top of the ridge are two gangly sassafras trees around sixty feet tall. They seem taller because of their position. Their shapes mirror each other, not-quite-identical twins, and all of their branches seem to grow out from the top third of the tree, as if they were limbed up. I guess it's the wind that keeps most of the trunk bare, but I don't know. I only know that beyond that ridge, the few scattered trees in the meadow have been shaped by their resistance to the wind.

The plant that gives the hillside its golden hue is broomsedge, so named because European settlers used the bristly dry stalks to make brooms. In summer, it's green, and it competes for attention with the rest of the meadow. But in the fall, its color changes to the golden orange that mesmerizes me. It grows to around three feet tall and stays upright when it's dormant, with a slight bend at the top like a stoop-shouldered old man. Broomsedge, or *Andropogon virginicus*, grows in bunches—it's not a sedge, despite its common name, but a bunchgrass like the other native grasses here. It starts out smooth near the base of the stalk and becomes more ragged as you move up. The *Andropogon* genus includes around 150 beard grasses, an apt descriptor. Broomsedge seeds are hidden in the frayed spots along the stem; the sparrow picks them out. The bobwhite wanders around with its head down amid the grasses, searching the ground for insects the way I used to search for money on the sidewalk when I was a struggling student. In spring, deer browse the young leaves, and this time of year they curl up in the meadow and leave car-size depressions in the grass.

From a distance you might think broomsedge is the only plant growing on this hillside, or the predominant one, but it's only the most colorful one right now. This field that slopes away below the sassafras trees holds one of the more pristine collections of native plants on the meadow: little bluestem, indiangrass, beaked panic grass, purple love grass, tick trefoil, native thistle, wild strawberry, goldenrod, dogbane. Besides the fescue, there are fewer nonnative plants here, and there is less of the native but overzealous blackberry that seems intent on crowding out everything else.

Despite a nearly 30 percent grade, this hill was cleared a long time ago, first to plant the apple orchard and later for cattle. But the accident of its topography is part of what makes the hillside's native plant life more diverse. Its soil remains highly erodible and acidic, lower in nutrients and best suited for the plants that evolved along with it. In turn, the creatures that most successfully use these plants for food and shelter are the ones that have evolved over thousands of years to do so. They're meant for each other, these plants, the insects, the birds, the organisms in the soil.

According to Virginia Working Landscapes, broomsedge is a "central component of native grasslands." Not just one component, a *central* component. Here in Virginia, it grows where it grows largely because it's supposed to. In a native meadow, the broomsedge won't take over; it's kept in check by other plants that share its affinity for the unenriched soil. It's drought tolerant like many of the natives here, and on this hill, in particular, it helps prevent erosion.

But there's a reason broomsedge has been called poverty grass. For farmers, it's a sign of depleted soil. The grazing process, if managed carefully, reinforces the nutrient cycle that the farmer prefers in a pasture; even so, fertilizers are often needed year after year to maintain nonnative grasses for cattle. As with crownbeard, cattle can eat broomsedge in late spring and early summer when it's young and green, but once it matures into that straw-like stuff, they won't eat it, and then it can take hold and spread. I had no idea cows were picky, but they are, and this may be a lifesaving quality, since a number of native plants toxic to cattle and other livestock might opportunistically grow in a field. (Native dogbane, which is related to milkweed, is one of these, and it's growing profusely in the

Two Ways of Looking at a Hillside

meadow right now, more widespread than it was only a year ago.) Sheep, on the other hand, are not picky enough. One farmer told me a sheep will eat something and drop dead right there in the field. Another put it more succinctly: "Sheep are always trying to die."

If I were raising cattle here, I might stare at that golden hillside in dismay rather than admiration at this now-inedible stuff that seems good for nothing but sweeping a dusty floor. If I thought of broomsedge as a weed that was keeping my animals from eating well, my first impulse would be to get rid of it and begin to take back my field.

When I search for information on broomsedge, most of the results point me to education-based sites that tell me it's an undesirable weed. A study out of Kansas State University examined when and how to fertilize a pasture to prevent a takeover of broomsedge. I could fertilize with lime and phosphorous, or cow manure ... and wait. It may take longer to revive the pasture this way, but it will work eventually. In the meantime, I'd have to supplement grazing with expensive hay until the pasture grass grows back. Only a few sites I found, like the University of Florida extension blog, prominently state that broomsedge is a native grass.

If I didn't know much about it, how would I know it's a native grass—and if I did know, would I care? When Europeans first arrived in Virginia, they brought their own seeds and their own animals with them; they planted grasses from back home that their livestock, also from back home, were accustomed to eating. If I raised livestock, how would I know that my cows could thrive eating native grasses without the need for fertilizers? When feed is the farmer's number-one expense, this could save me a lot of money in the long run. But those toxic plants, even the "good" ones, would still need to be banished.

Broomsedge produces seeds later in the summer, and if I want to prevent that, I need to kill the plant before it happens. That might mean using glyphosate, the chemical in Roundup, and broadcast-spraying herbicide around acres of field from a boom up to one hundred feet long that sits on a truck with a tank holding as much as fifteen hundred gallons of chemical. But, glyphosate is nonselective and as such would kill the feed grasses and forbs a farmer might want to retain. One option is a brush treatment, in which a truck drives through a field and brushes herbicide only on plants that are above a certain height. The Zabulon skipper, a small yellow and brown relative of the butterfly, lays its eggs on broomsedge and other native grasses here; its larvae hatch and feed on the grasses. Herbicides can cause collateral damage, threatening the birds that rely on broomsedge for food and shelter, the bees that pollinate everything—both wild bees and that introduced agricultural tool, the honey bee—and any insects that happen to be in the wrong place at the wrong time. I could graze my cattle without thinking about the Zabulon skipper, but I don't have any cattle. So I'll think about the skipper, every time I see that hillside's golden flank.

***

My friend Adams is one of the first people I got to know here. In addition to running a bush-hogging operation and an excavating operation, he raises beef cattle. He's a proud dad, showing me photos of his toddler who shares his mischievous grin. Adams is full of helpful advice about farm equipment—which side-by-side can handle these hills, and which brush cutter we want for the job of carving away bramble—and about the general hazards of country living. He takes every opportunity to remind me that the brush cutter is dangerous. Maybe it's obvious that I'm a klutz? Adams's cautionary tales always seem to end with something terrible happening, like the woman who ended up in intensive care after she pulled on a clean pair of trousers and was bitten by a brown recluse spider that had hidden inside them. This time he tells me about two men who were operating brush cutters together in a field, when one of them got his leg torn open by the blade the other one was wielding. *He died.*

I hold the machine gingerly and practice with the string trimmer.

It's May, and we're talking about bush-hogging a new path around one of the fields, when Adams tells me he thinks I should let him cut back the plants growing alongside the last quarter mile of the gravel road

Two Ways of Looking at a Hillside

leading up the mountain. I say that a lot of milkweed grows along that part of the road. He thinks I'm complaining.

"I hate that stuff, too," he says.

Milkweed is toxic to livestock, and as such, he can't let it grow freely in his pasture. Cows may be picky and good at avoiding the plant when they're grazing, but it can be dangerous when it's dried and mixed in with hay where it won't be noticed. But what about the monarch butterfly that can lay its eggs only on that plant?

The monarch is the one butterfly most of us were taught to recognize from the time we were kids. Thirty years after I learned about it in school, my kids were studying it, too. As for milkweed, most of us suburban kids had never seen it in the wild; the teacher brought in a puffy seed pod to pass around so that we could all get a close look.

The monarch is one of only a few insects that migrate, and it migrates in two directions, like a bird. The eastern monarch travels from Canada to Mexico, where it overwinters in the forests of Michoacán before returning north. It takes as many as four or five generations to make the round trip. The western monarch winters in California. Over twenty-five years ago, fresh off a hike to see the elephant seals gathering on the beach at Año Nuevo State Park, after watching the males occasionally bash into each other like opposing football linemen, I drove to the Monterey town of Pacific Grove to witness a more peaceful natural display. The butterflies were gathered in the thousands, packed together on the branches of Monterey pines for warmth. There were so many it almost didn't seem real, and soon it may not be. Back when I visited Pacific Grove in the late 1990s, there were more than a million of the butterflies counted; by the latest count that number had dropped to around 230,000.

On the trip from south to north, the eastern monarch lays its eggs on milkweed plants. When the caterpillars hatch, they arrive on their only food source in the world. Like 90 percent of the world's insects, the monarch caterpillar can only eat one type of plant. Similar to the Very Hungry Caterpillar in the much-loved book of the same name, it can't eat anything besides that green leaf—in this case, only the milkweed leaf—without becoming ill. Unlike the fictional caterpillar, the real-life insect knows better than to try. The monarch caterpillar is immune to the milkweed's toxins, defenses that otherwise protect the plant from being dinner for all but a few species of beetle. These same toxins make the caterpillar an unpalatable meal for most predators as well.

As a kid, I remember seeing monarchs around every year, even if I didn't see milkweed plants. But I hadn't spotted one in the wild for many years until several of them casually fluttered by me on the mountain. If this is the one butterfly we're all taught to recognize, why don't we see more of them? Because, since the 1970s when I was a kid, they've decreased in number by *96 percent*.

Climate change and logging have impacted the eastern monarch's wintering grounds in Mexico, and development and pesticides used to kill milkweed when it grows in crop and hayfields have decimated the caterpillar's single food source. In the Midwest, where the highest concentration of north-migrating monarchs arrive looking to lay eggs, glyphosate-resistant GMO corn and soybean fields allow farmers to use the chemical to kill unwanted plants, like milkweed, that grow up among their crops, without harming the crop plants. Milkweed often proliferates around the edges of farmers' fields, and its well-developed defenses are no match for repeated applications of pesticide. As they reach the Midwest, the tired butterflies have trouble finding enough milkweed plants where they can lay their eggs.

Growing up, I thought milkweed, or *Asclepias syriaca*, was rare. The only time I saw it live was in a botanical garden. Imagine, then, my excitement when I discovered milkweed on the mountain. It's one of the dominant native plants on this land, and wandering the meadow, you wouldn't know that it's scarce in other parts of the country. I even find some of the sun-loving plant growing in the deep shade of a poplar tree. With all of the milkweed here, along with profuse goldenrod and aster, where the adult monarch likes to collect nectar as sustenance for its long journey, I don't understand why I haven't seen more monarchs. But I can't control what happens along their migration route before they arrive here.

Two Ways of Looking at a Hillside

Last year, I picked a few seedpods from the milkweed on the mountain and brought it to my suburban street. I offered seeds to neighbors to spread in their gardens, and I did the same in the few available sunny spots in my yard. The following year, nothing came up. I chalked it up to yet another failure—I was still a not-gardener, and planting seeds was gardening, even if they were seeds I collected in the wild. I'd somehow fumbled it. And, although we'd stopped fertilizing years ago, I thought it was possible that the soil around my house had been overtreated for too long, making it inhospitable ground for milkweed.

You can buy milkweed plants in a nursery and plant them in your yard, and well-meaning people have done so in the hope of rewilding their yards—returning them to nature—for the benefit of native pollinators. What they might not realize is that plants sold at nurseries may have been sprayed with pesticides and fungicides that can harm the beneficial insects they're trying to save—the wild bees that pollinate the flowers, and the caterpillars that eat the leaves. A recent study by University of Nevada scientists collaborating with the Xerces Society examined more than two hundred milkweed plants sold by more than thirty nurseries in fifteen states. An average of twelve pesticides were discovered on every plant, with a total of sixty-one different pesticides, only nine of which had been studied for their impact on monarchs. More than a third of the chemicals were found at levels higher than an amount deemed to have a sublethal impact on the insects. Sublethal impact could include anything from impaired navigation to failure of eggs to hatch to shortened life spans—and what does *higher* than sublethal mean, but *lethal*? The size of the nursery where the plant was sold and the variety of milkweed sold didn't matter. I could be sure that the seed pods I'd collected had never been sprayed because I hadn't sprayed the fields where they were growing. But, the study concluded, even "plants with labels advertising their value for wildlife did not have fewer pesticides at concentrations known to have a negative effect on monarchs." It never occurred to me to wonder whether nursery plants are treated with pesticides, and I'm sure I'm not alone. (If you're trying to do a good thing by planting natives—and that is a good thing!—or planting anything, for that matter, the only way to make sure you're not inadvertently making things worse is to ask the nursery some probing questions before buying *any* plant.)

This spring, I was surprised and excited to see milkweed growing right where I'd spread the seeds more than a year before. Maybe they needed that extra year to get used to the new locale; I'm just happy I could bring a piece of the mountain to suburbia, and maybe provide a new stopping point on the monarch's long journey. But most livestock farmers, like Adams, will experience a different emotion on finding milkweed growing near their fields. At this point, Adams was still under the impression that I was hampered by wrong assumptions and inexperience (some of which was certainly true), rather than different goals. Good thing we can laugh about our differences.

"Don't you worry," says Adams, about the plants along the gravel road. "I'll spray that stuff with 2,4-D."

"No, please don't spray the milkweed!" I say. "It's important for the butterflies!"

Adams chuckles and shakes his head, as if I'm totally naive and destined to learn the hard way. And maybe I am. "All right," he says, "but I'm telling you, it's just gonna keep on growing."

I'm okay with that.

There's obviously an inherent conflict between many of the native plants that grow in Virginia and elsewhere, and the way people have used their land for years and sometimes centuries. Milkweed, crownbeard, and broomsedge are three of the more prevalent native plants on the mountain and among the first plants I learned to identify here. Yes, it's partly because they can spread easily in the right circumstances, and they're tough enough to compete with nonnative grasses, but they don't have to be shy, retiring flowers to be an overall plus. The way to save native plants, and save the insects and birds that need them, is to let them grow where you can, even if it means around the edges of farm fields. While I work on restoring the meadows, these plants and others like them are among those I'll be looking for as signs of the improving health of the land.

Two Ways of Looking at a Hillside

Whenever someone in my family gives a gift, they say "Use it in good health." The automatic utterance may have originated as a way to ward off bad luck; now, when I find a new patch of broomsedge or milkweed thriving on the mountain, I try to project the same protective spell onto the meadow, to *use it in good health*.

I'm not sure what it is about the humble broomsedge that calms me—its fall glow? The way it appears windswept even on a still day? I only know that when I see it, I feel optimistic about what I'm hoping to accomplish here. I feel like things might work out. Maybe because, unlike so many other plants I've slowly learned to recognize, the broomsedge is supposed to be here. It belongs. Maybe I belong, too.

Chapter 7

WE BROUGHT THIS ON OURSELVES

... in which a wildflower becomes a weed, weeds threaten to engulf the planet, and the original invasive species is ... me

When I started this project, I had, I admit, a rather pie-in-the-sky view of how I'd reshape the mountain into a native plant paradise, a model of conservation, a meadow that would inspire Julie Andrews to break into song. Since then I have, as they say, gone through some things. I've met with a healthy dose of realism. It may be reasonable to expect to eliminate unwanted plants on a small piece of land the scale of a suburban backyard, but on many wild acres of field and forest, it's like trying to carry water in a bucket full of holes. I no sooner weed around one boulder than I notice the same weed sprouting up in a new spot on the other end of the meadow, shielding itself cunningly amid native goldenrod, aster,

We Brought This on Ourselves

and wild bergamot. It's tricky to extract only the offending plant, and even trickier to spot treat it without harming anything else.

I'd decided that until or unless a funding agreement *required* me to spray large areas with chemicals, I'd manage as much as possible by pulling or spot spraying or target-treating invasive weeds. That fit my budget and my philosophy; I wouldn't undertake the costs or risks of broader spraying without assistance. That would also give me time to learn more about my options.

Enter spotted knapweed, officially known as *Centaurea stoebe* subspecies *micranthos*. (Say that three times fast.) Spotted knapweed is an invasive plant with fringey pink flowers that remind me of my grandmother's old-fashioned bathing cap, and a bract, between the flower and the stem, that resembles a pineapple. The plant can grow as tall as five feet; its seeds are released by the slightest touch to fly up to three feet away, and they're transported even farther by wind, passing animals, farm equipment, and people. A single square foot of the weed can produce anywhere from four thousand to more than forty thousand seeds, and those seeds remain viable for at least eight years. Plus, it's a perennial, so like sequels to the movie *Halloween*, it will keep coming back. If you try repeated mowing, thinking that will exhaust the plant—nope, it will only learn to seed closer to the ground. Still, ecologist Ken Thompson, in his book *Where Do Camels Belong? Why Invasive Species Aren't All Bad*, concludes that spotted knapweed isn't especially harmful to its native neighbors. I find this doubtful given that, on top of its prolific seeding and its ability to invade even undisturbed fields where native bunchgrasses are growing—as it's doing in parts of this meadow—the plant is thought to release enough of a chemical called catechin into the ground to repel other plants that try to grow nearby. That's because the plant is *allelopathic*, a trait shared with some other invasive weeds and even some native plants, like black walnut trees, goldenrod, and sunflowers. Spotted knapweed, like the sunflower, is part of the aster family—clearly the branch that's no longer on speaking terms. Allelopathic plants release toxic chemicals into the soil to suppress the growth of other plants and reduce competition. These chemicals can repel, weaken, or even kill other plants, alter soil organisms, or interrupt the normal behavior of the soil fungi that help native plants grow. Some plants release toxins through their roots, whereas others release them through dropped leaves as they decay; some will do both. And the toxins can continue to act in the soil for varying periods even after the offending plants have been removed. Some plants are more sensitive to these toxins than others, and some plants' toxins are more effective than others. Along with its prolific seeding, spotted knapweed's allelopathic tendency helps it take over a field and form a monoculture, a homogenous zone that excludes other plants. These monocultures can lead to population declines and extinctions among local plants and wildlife.

In Montana, spotted knapweed now covers large areas of rangeland, and in those areas, elk habitat has decreased by 98 percent. Because elk can't eat spotted knapweed, the plant has significantly reduced their available food supply, and every year there are hundreds of losses to the herd as a result. According to the Thurston County Noxious Weed Sheet, the best treatment is to not let the plant grow in the first place. "Above all," it advises, "prevent plants from going to seed." Here on the mountain, it's a little late for that.

* * *

Invasive species are a worldwide problem. The Intergovernmental Science-Policy Platform on Biodiversity and Ecosystem Services (IPBES) was formed in 2012 by ninety-four countries, including the United States, to work on global conservation policy. According to a 2023 IPBES report, invasive species "play a key role in 60 percent of global plant and animal extinctions, and cost humanity more than $400 billion a year—an amount that has quadrupled every decade since 1970."

North America is losing close to a million and a half acres of native plants to invasive plants every year. As the Virginia Department of Conservation and Recreation (DCR) asserts on its website, "An invasive plant infestation is like a slow-motion explosion." These plants are both a threat

We Brought This on Ourselves

to the stability of ecosystems and a threat to biodiversity. They reduce, and in many cases eliminate, food and shelter available to the wildlife that call a place home. Native creatures like those unfortunate elk depend on native plants for their existence, and when those plants are threatened or pushed out, they have trouble reproducing and successfully raising and feeding their young.

Nonnative plants are undesirable even if they're not invasive, because they take up space that could be filled by native plants without replacing their contributions to the soil or the food web. Conservation biologist Desiree Narango, working with Douglas Tallamy, entomologist and author of *Nature's Best Hope*, examined suburban backyards and found that the chickadee population could only be sustained in places where no more than 30 percent of plant matter was nonnative. This was the first study to look at the correlation between the decisions we make about our backyards and the impact of those decisions on bird breeding success. Nonnative plants aren't automatically a severe ecosystem threat, especially when their proportion in the landscape remains low, and there are plenty such plants that don't become invasive—but when new plants are introduced, it can be next to impossible to predict which ones will eventually cause a problem.

What makes a plant *invasive*, then? Introduced plants that become invasive are adaptable: they can establish and survive in a range of climates and a variety of habitats with differing soil quality and levels of sunlight. They're fast growing and fast spreading, usually capable of reproducing in multiple ways. They're prolific seeders, and their seeds have a high viability rate (meaning, a high percentage of the seeds can produce a plant). They're quick to fill open areas and are often the first plants to begin growing in a place that's been recently disturbed by human activity. These fast-growing pioneer plants gain an advantage over slower-moving native plants, and, once established, they can prevent, or inhibit, the return of native plants.

Allelopathic invasive plants like spotted knapweed make the soil unusable for native plants. Some of these plants will even turn mycorrhizal fungi's ability to provide nutrients to their own advantage. (These fungi normally help feed plants and can serve as nutrient paths linking different species of native trees.) And, finally, invasives don't support native wildlife as food or shelter at any level, or at a level that makes up for their negative impacts, like the reduction or loss of the native species their success imperils.

The introduction of nonnative plants started as soon as a nonnative person set foot in North America with a pocketful of wheat and barley seeds ready to plant in the new land. The cherries and apples that grow here on the mountain came from elsewhere. European settlers in the 1600s brought not only crop plants but, unknowingly, seeds for weeds like dandelions and crabgrass. In the mid-eighteenth century, horticulturalists and plant-enthusiast landowners sought new and exciting exotic ornamental plants to display in their gardens. International trade in plant species here started with John Bartram, founder of the Philadelphia botanical gardens and official botanist to King George III. Thomas Jefferson, George Washington, and James Madison were among those who obtained exotic seeds and plants from traders in Philadelphia and planted them on their estates in Virginia. Madison's journals refer to at least twenty imported plants from seven different countries, including rice from East Timor and black-eyed peas from France. But these plant introductions weren't just a fun horticultural hobby; they were a tool of colonization and a power play in which landowners claimed and remade the landscape as they saw fit. The introduced plants displaced native species, while the colonists pushed out native people.

In the late nineteenth century, the USDA established a foreign seed and plant program that sent explorers around the world—including parts of Asia, where many of the invasive plants I'm struggling with originated—making exotic plant introduction official policy. Soon the plants began to show up in nurseries that sold to the public, where they became wildly popular. More than half of the invasive plant species in the United States, including Japanese knotweed, poison hemlock, buckthorn, tree

We Brought This on Ourselves

of heaven, giant hogweed, and Japanese barberry, were introduced this way, as ornamental plants that became too "successful," escaping the confining walls and fences of private gardens and making their way into the wild. Some scientists place that number as high as 85 percent.

The plants came surreptitiously, as stowaways by ship, train, or car. They came incidentally, used as packing material to protect fragile items sent from abroad, or as seeds hiding in nursery stock. It's thought that spotted knapweed arrived from Central Europe in the late 1800s in a contaminated shipment of seeds and in soil used as ballast. While you're reading this, invasive plants are probably being purchased from nurseries and home supply stores that still market them as "hardy" ornamentals, even though in some states they're classified as noxious weeds.

As of about ten years ago, there were more than thirteen hundred plants in the United States identified as invasive; a recent count isn't available, but that number has only increased. The USDA's National Invasive Species Information Center admits, "The large numbers ... prevent us from maintaining detailed information on *all* invasive species." As of 2017, invasive plants had infiltrated nearly a million and a half acres of national parkland, but only about forty-three thousand acres are at a level that staff can keep under control. The National Park Service has suggested that the threat invasive species pose to biodiversity is second only to habitat loss. I don't see those as separate problems—invasive plants lead to and exacerbate habitat loss, and they're often boosted by it as well; these threats are as intertwined as Asiatic bittersweet around a tulip poplar.

Habitat loss is a feature of human development. Any activity that precipitates upheaval in a natural corridor creates an opening for invasive plants to spread and thrive. These plants are ubiquitous in areas disturbed by humans—housing, commercial parks, roads, bridges, train tracks, the outskirts of tilled fields. When the power company sprays to clear vegetation under power lines, openings are created for invasive plants. All along the gravel road that goes up the mountain to the meadows, invasive stiltgrass has replaced native flowers. How do I know spraying was one probable factor? Because the upper section of the road, where there were no power lines, was full of native milkweed, goldenrod, and aster, until an unfortunately necessary disturbance—roadwork—led to an outbreak of yet another weed.

In the mountain meadows, haying introduced seeds that didn't belong. Tractors and mowers unintentionally dropped seed along mowed paths from one end of the mountaintop to the other. To effectively remove unwanted seeds hiding in the hundreds of crevices in farm equipment would require a dauntingly thorough and time-consuming cleaning. If you live in a place where leaf blowers are the norm, when you blow your leaves to the curb, whatever is seeding on your land is being airmailed to your neighbors. When your county clears leaves or debris from the highway right-of-way, seeds from weeds growing at the edge of the road are blown into the woods.

Even well-intentioned hiking and camping helps nonnatives spread. Unless you clean your boots and equipment every time you're about to enter a park, you're carrying whatever you came into contact with before you got there. Your dog is a walking seed-and-plant-distribution system. After a day on the mountain, I spent over an hour combing out tiny round seeds that had cocooned themselves into my dog's coat. (Note to self: give this poodle a crew cut.) Of course, wild animals, especially deer, help spread those seeds, too. But the seeds wouldn't be there in the first place if we hadn't introduced them, and they'd have less opportunity to spread if we weren't constantly disturbing natural areas. It's on us to keep the problem from getting worse.

In Shenandoah National Park, the jewel of the northern Blue Ridge, there are 350 nonnative plants, 25 percent of the plant total in the park. Ten percent of these plants are considered invasive, and they're found in more than one hundred locations. The trouble is it can be far easier for the plants to expand their presence than it is to make a dent in their numbers. They'll spread if we don't do anything, or don't do the right things, or don't do enough. My own experience should serve as a lamentable example.

We Brought This on Ourselves

***

Early on, when we were still figuring out how crazy it would be to buy the land (answer: extremely, totally, hard to overstate), we asked a forester for his opinion about the condition of the forests here. As we approached the wood's edge, he stopped and pointed far across the meadow and delivered a warning.

"See that bush with the silvery leaves?" There was a single unassuming-looking bush about three hundred feet away on the far side of the overgrown field. "Make sure you take that out. Any bush that looks like that." The autumn olive is the only silvery green shrub amid the deep greens and browns of the meadow plants. Its bark and even its berries are speckled with silver. There were only one or two in the meadow, but they'd take over, he warned. He'd seen it happen.

The autumn olive (*Elaeagnus umbellata*) is no relation to the olive tree. It first arrived from Asia in 1830 as part of the exotic plant trade. Later, the USDA's Soil Conservation Service (SCS), the precursor to today's NRCS, decided the shrub could be used to control erosion, and in the 1940s began marketing it through its district offices for soil control along roads and around bridges, and as a beneficial food plant for wildlife. Farmers were advised to use it for hedgerows to create windbreaks around crop fields, or in dry, highly erodible areas to prevent topsoil from blowing away, and it was planted over former mines. The autumn olive was planted intentionally all over eastern and central states until the 1970s; Virginia's own Department of Forestry planted it throughout the state, as a DOF forester sheepishly admitted to me. The particular cultivar the SCS chose to promote was the Cardinal, selected for its larger and more numerous fruits—it produces more seeds than other autumn olive varieties—and because it can grow as much as six feet *in a single season*.

The shrub tolerates salt and drought. Wherever it grows, it alters the soil composition, making it inhospitable to native plants and organisms and better for other invasives, which can handle, and often prefer, higher levels of nitrogen in the soil. That's why some invasives can do especially well in and around crop and hayfields, and pastures that have been fertilized—the nitrogen level is higher than what most native plants here are accustomed to. But the autumn olive doesn't need to find a fertilized field (although it will be happy to grow there, too); it does the fertilizing itself by holding higher levels of nitrogen in the soil, in a process called *nitrogen fixation*. Worse yet, the excess nitrogen the plant emits can leach into groundwater and pollute streams and rivers, leading to harmful algal blooms that can result in dead zones where there isn't enough oxygen to support life. Some of these algal blooms are toxic, and those toxins are passed on to anyone who drinks from the water or dines on anything contaminated by that water, including plants or animal life.

The prolific autumn olive produces thousands of berries, which in turn result in tens of thousands of seeds from a single bush each year. The shrub's distinctive leaves emerge early in spring, and because of its wide, bulbous shape, it shades out native plants that emerge later. New bushes sprout quickly and profusely from those many seeds, so wherever you see adult plants, you'll find lots of seedlings nearby.

This shrub is only one in a string of now-invasive plants that were once introduced as beneficial. Thorny multiflora rose bushes were planted in the 1920s as fencerows around cattle pastures. Kudzu was distributed to farmers by the SCS for planting as a cover crop for erosion control and to enrich topsoil in the South, where soils had been depleted by poor management and overplanting of cotton crops. To sweeten the deal, the SCS actually *paid farmers for every acre of kudzu they planted*. It was a bargain—for the kudzu. Deer will sometimes nibble on it if none of their native favorites are available, but the plant is of low nutritional value and not a good substitute. The idea that cows and other animals would eat kudzu, and their nitrogen-filled poop would improve the soil ... did not go as planned.

Now it's out of control. It's as if we created a novel ecosystem in the midst of our native one, a real-time experiment in world-building, partly intentional and partly an accidental result of human interference. We

We Brought This on Ourselves

didn't understand the forces we'd brought together and unleashed. They escaped the boundaries we'd set, only to wreak havoc—*Jurassic Park*, starring plants.

Native birds and animals are willing to eat the sweet fruit of the autumn olive, which of course helps spread its seeds. Humans can eat it, too; I've heard of people making jam from the berries. A book on foraging in the wild instructs you to loosen the berries with your fingers and plop them into a bucket, or shake them onto a tarp to collect them. It also describes the bush as "12–18 inches tall." I've never seen an autumn olive that tiny. Is it because they grow so quickly, or is it that I can't see them amid the tall grasses and bramble until they're at least three feet high? Either way, I doubt they stay that size for long. The book calls it a "delicious, invasive plant" and suggests that, as such, there's no need to limit harvesting. Just watch where you toss those seeds.

Sure, if you have fruit, you may as well use it—right before you cut the bush down and brush the stump with herbicide so it won't grow back. But you won't beat the birds or the bears or the deer to all of the berries on all of the bushes.

If native birds and wildlife will eat the fruit, maybe there's no harm? Maybe it's actually an ecosystem plus, right? As recently as twenty years ago, the USDA studied the autumn olive as a potential alternative crop because of the berry's high lycopene content. (The study doesn't acknowledge its invasive tendency, calling it "vigorous," which is some euphemism for "unstoppable.") But for birds and bears, eating the fruit of the autumn olive is the same as if they'd traded their healthy lunch for a Twinkie—the berries are full of empty calories, and the animals and their young will suffer.

Say when your kids are hungry, you hunt all over for fat caterpillars to regurgitate into their mouths, but you can't find enough in your usual territory. The elderberry and redbud that used to support those caterpillars have been crowded out by autumn olives. Most butterflies and moths will only lay their eggs on plants their larvae can eat. They need a meal as soon as they hatch, and many can eat only one type of plant. You won't find butterfly eggs on the autumn olive—the larvae can't eat it. If their host plant is missing from the environment, those insects won't hang around. In this scarcity situation, you, the bird parent, must choose among objectionable options: feed your brood less often or feed them smaller insects that don't provide as much protein. This could lead to a failure to thrive or even starvation. The other option is to abandon the territory and nest elsewhere, and those "elsewheres" are growing scarce. The chickadee study showed that when faced with these lower-value food choices, birds may rear fewer nestlings and fewer broods, and fewer of these nestlings will make it to fledge. When invasive plants make inroads, the impact is broad and deep.

The autumn olive now grows from Maine to Florida and has been found as far west as the Pacific Northwest. It's on the federal noxious plant list, as well as similar lists in several states, and yet it didn't take me long to find nurseries in Washington State and Georgia that sell it. A source in Portland, Oregon, proudly claims to ship "to all 50 states." At least the latter mentions that the plant is invasive "in some regions." The trouble is, every state has its own system. Some invasive plant lists carry the force of law, and others don't. Some states don't keep a list at all, and if they do, the autumn olive isn't always on it. It's classed as highly invasive on the Virginia invasive plant species list assembled by the Virginia DCR, but that list is purely informational and has "no regulatory authority." In Delaware, the bush is listed as invasive, and it can't be legally bought, sold, or propagated there. In Maryland, the bush has been "under assessment" since 2017, and as of 2023, it was still being assessed. If it's placed on a list, depending on how it's classified, it could either be banned from sale in nurseries or permitted in nurseries, but with signage to warn customers about its invasive tendencies. While the Maryland Department of Agriculture was busy assessing the autumn olive, the Maryland Department of Natural Resources (DNR) offered a recipe for preparing autumn olive jam, encouraging citizens to cook and eat the berries before birds can get to them and spread the seeds. I don't know about you, but I've never gone to the Maryland DNR website looking for recipes; maybe I'll start.

We Brought This on Ourselves

***

To the extent that the harm caused by invasive plants is at all controversial, a small contingent of scientists argues that invasive species shouldn't be considered harmful without careful study, and that sometimes they can have a positive impact. This may sound reasonable on its face, but it's a position held by a tiny minority because so many species *are* harmful when they become established outside of their home continent, and they become harmful too quickly for field research to keep up. The chestnut blight wiped out chestnut trees; the emerald ash borer is currently wiping out ash trees; and then there's kudzu, which devoured an entire region of the country. A wait-and-see approach has not proved successful historically. Daniel Simberloff called the argument that invasive plants might not be problematic "a phony controversy," when speaking in the *New York Times* about the issue. Simberloff, a biologist and ecologist who was one of the early analysts of the theory of island biogeography before he came to specialize in the impact of invasive species, adds that this handful of contrarians has "vastly played down the severe harm that alien species caused." In one study, he notes that species transplanted from another continent are *forty times* more likely to become invasive than native plants that are moved within a continent.

Spotted knapweed, for its part, is unfazed. A special feature of invasive plants is that they're expensive and difficult to remove, and the larger the area, the more expensive and difficult. It might feel less overwhelming if I were working on a small suburban yard, along the lines of the burgeoning Homegrown National Park movement, which was founded by Douglas Tallamy to encourage homeowners to replace their lawns with plants that support a range of native pollinators. But working on a scale like this mountain, the challenge is exponential.

I'm sometimes able to extract the spotted knapweed's frustratingly deep taproot in one piece with my weed wrench. But can you imagine how long it would take to bushwhack through two acres, with a twelve-pound weed wrench slung over your shoulder so it doesn't get hung up in the bramble, searching for a specific weed amid a mass of random overgrown plants? Now try it with ten acres. Now twenty. When you're done with those twenty, you'll find knapweed has somehow migrated to another five- or ten-acre field where you hadn't noticed it before. Oh, and wear gloves, because spotted knapweed has been known to cause a rash. Plus, a chemical it emits may cause cancer. And mind the rattlesnakes. Happy weeding!

The website of the DCR, with its inimitable dry wit, says that "management of invasive species can appear to be a complicated and unending task." When I started this project, I thought I'd restore the meadow *a little at a time.* Which is fine and practical—until I realized that the many acres that are waiting for me to get around to them are not lying there dormant, frozen in time. When my kids were young and went to sleep-away camp, they seemed upset to hear about anything that happened in their absence, from a dinner out to the demise of an old vacuum cleaner. In their minds, we'd lapsed into a coma the moment we dropped them off and stayed that way until they remembered us because they needed us again, like when one of them mailed three consecutive letters in three days that read only, "Chapstick! Please send Chapstick!"

The human body sheds one million skin cells daily, and change is occurring in every quadrant of every field every day in every season, and at times I feel helpless at the task in front of me. I may have Celia Vuocolo's plan describing the steps I should take—spraying, burning, removing weeds, spraying, and burning again—but how to approach those, when I can't do it all at once? The where and how are open to interpretation and serendipity. And everything seems to come around to invasive plants.

It's like playing a game of tag, and catching the one who's "it," only to find there are a dozen more kids who are "it," and they've scattered in as many directions. You can never catch them all. I'm conveniently overlooking the rest of that metaphor. In an actual game of tag, after tagging whoever was "it," *I'd* become "it." And that fits. I'm the worst invasive of all. I may not be a plant, but the plants themselves aren't to blame for what's happening.

We Brought This on Ourselves

I can't imagine maintaining the autumn olive as a hedgerow as the government once advised, when its natural inclination is to expand into a thicket and keep expanding, like a plant version of the Blob. That's exactly where things were headed on one of my steepest hillsides, south of the big oak tree, an area too steep and overgrown with bramble for me to easily access it. Some "shrub": autumn olives can grow to twenty feet tall and thirty feet wide. The bushes were too large and numerous for me to pull them up or cut them down myself. I added it to Brian Morse's to-do list to treat along with the invasive tree of heaven (*Ailanthus altissima*) and the paulownia (*Paulownia tomentosa*, which sounds like "torment" for a good reason). I think of these as the triumvirate of trash trees, and I'm desperate to keep them from further impinging on the meadow from the edges, from the middle, from anywhere. I asked Brian to save work on the slope below the oak for last, on the theory that, as the place farthest from the central meadow, it doesn't pose as imminent a threat. Plus, its steepness makes it even more complicated and daunting to work on than the rest of the fields.

When the first forester had advised us to get rid of "that bush," I figured that naturally he meant we should cut it down. (*Isn't that what foresters do?* I thought. *Cut things down?* I didn't know anything.) I especially didn't know then that you can't merely cut down a woody invasive unless you want it to return with a vengeance, and autumn olive is no different— or it *is* different in that it's worse than some others. If you cut it or mow it and don't treat the stumps, it will resprout, and the new growth will be stronger than before, with denser, harder branches that are even more difficult to cut.

But Brian ran out of time before he got to the oak slope, and so did my budget. I'd have to find another way to manage that hillside. In one of the many ironies in my attempts to deal with invasive plants, my efforts to slow the expansion of the autumn olives had, let's say, unintended consequences. I'm beginning to think that would be a good name for this place: Unintended Consequences Farm.

Chapter 8

UNINTENDED CONSEQUENCES FARM

... in which an invasive vine eats my brain

I'm constantly shifting focus and emphasis as the landscape shifts under my feet. How do I address even a handful of the umpteen problems I've had to ignore while dealing with the ones that seem most urgent and time sensitive, or those that have the longest timeline so I'd better get started if I want to spur change in my lifetime? For the first year, the autumn olives on that hillside would have to wait while I dealt with pressing needs elsewhere. But they didn't wait, of course. They kept growing and fruiting and spreading, and creatures ate the fruit and transported seeds farther afield, and finally I had to do *something*.

For the uninitiated, which until recently included me, there is mowing, bush-hogging, and forestry mulching. Mowing is for relatively short, weak, herbaceous vegetation; bush-hogging is meant to cut out overgrown vegetation that's taller or tougher than a regular mower or a zero-turn can handle. Forestry mulching is for anything too dense and woody for a bush hog—that is, woody growth that's more than a few inches thick. Bush hogs and forestry mulchers come in different sizes, depending on the size and volume of what you're trying to cut. If you have the right equipment, you can knock over one of those giant autumn olives, root ball and all, and chop it to pieces so there's almost nothing left. The mulcher scatters the shredded wood over the ground in a way that should allow it to break down into the soil and help prevent erosion. The quick and dirty solution to the autumn olives was forestry mulching. It was really a stopgap measure until I could do something more "permanent."

But I hesitated to clear that hillside below the oak, not only because I was ignoring it for the other seventy acres of meadow, but because, like so much else, the advice I'd received about mowing was contradictory (and by *mowing*, I mean any of the methods I described). My expert advisers occasionally express conflicting opinions and perspectives, based on their professional background and orientation; sometimes these differences are subtle, and sometimes they're stark. Those varied perspectives taught me that this work isn't cut and dry; there's no such thing as one-size-fits-all in clothing or in native meadow restoration. And one of those points of disagreement is whether or not to mow. Celia had advised me not to use mowing as a lone management tool for good reasons. But maybe this hillside was a unique problem that warranted an exception. I should have known that my search for answers would only lead to more uncertainty.

Adams, for his part, thought I'd regret not keeping *all* of the fields mowed. "If I was you," he said, "I'd mow the whole dang thing." If I didn't, he warned, waving at the crownbeard and dogbane, all the wild stuff would take over. I told him about the cycle of spraying and burning that had been recommended for promoting native grasses, instead of mowing, which had made it harder for those grasses to grow.

"You don't have to burn anything," he said. "Just keep it mowed, and you'll have grass."

We were obviously talking about two different kinds of grass, which made perfect sense, since, like most cattle farmers, Adams grazes his livestock on nonnative pasture grasses. I told him I wanted the wild stuff, and I started to list the native grasses I was trying to encourage: little bluestem, broomsedge, indiangrass...

He latched onto broomsedge. "I can't pull that stuff out fast enough," he said. "It means the soil is sour."

That's when I got on my hobbyhorse, droning on about the importance of native grasses and a native plant ecosystem, and about the plan I was trying to follow. I might have come off as a tad ... overzealous ...

Adams turned to B. "Is she mad at me?"

He reassured Adams that I get like that when I'm fixated on something.

I was irritated with myself for overreacting. At times it felt difficult to contain all of the information and advice I'd collected, as if I'd crammed for a test, but the only examiner was me. My enthusiasm could be annoying, even to B. (um, especially to him; he was on my side in most of my decisions about the fields, but he didn't see it as quite the emergency I did). And the frustration of being unable to control what was happening with all of these plants was getting to me—I was used to being "in charge" of my environment. A laughable idea.

Although most everyone in the ecology community considered regular mowing no substitute for conservation land management, one or two did acknowledge that occasionally mowing some fields I wasn't able to work on, if done at the recommended time (in March, before spring growth begins and at the end of the winter hibernation period for most wildlife), wasn't out of the question as a measure to try to slow succession and check invasives. Aha! I pounced on this as permission to mow. The lesson is, if you ask enough people, you'll eventually get the answer you were hoping for.

The monstrous autumn olives on that hillside were advancing inexorably toward the upper meadow. The problem couldn't wait any longer. In

Unintended Consequences Farm

March, when I finally gave the okay for Adams to take his skid steer and forestry mulcher to that hillside and those bushes, he was more than ready. But I couldn't expect him to stop and get out of his truck to treat each and every stump he cut down. Plus, Adams had told me on many occasions that he was absolutely not stepping foot off his equipment in those fields where the rattlesnakes are plentiful, and that I was a fool to do so myself. (One of the names I jokingly called the place, Fool on the Hill Farm, was too apt at times.) He showed me the pistol he kept on the front seat of his truck; it was good for unexpected encounters with ornery reptiles, among other dangers. I thought of Edward Abbey, who said he'd sooner shoot a man than a rattlesnake.

The herbicide treatment has to happen right away, because, within minutes, the bush raises its shields, like a Klingon spacecraft. After that, nothing will penetrate, rendering the chemical useless. I thought of following along behind Adams' skid steer and applying the chemicals myself, but that seemed unsafe for a number of reasons, not only because I hadn't yet found a pair of snake-proof boots that fit me. (At times it felt like my best option was to hide in a dark room until I forgot all about the autumn olives on the hillside.)

I knew by not treating the stumps I could be risking regrowth. But Adams assured me he had a solution—he didn't plan to leave any stumps. I wasn't sure if this was just bluster, but it turned out it was true; there would have been nothing there for me to treat. With the forestry mulcher, he removed even the most massive bushes. And when I say removed, I mean pulverized. Afterward, I picked my way down the slope, which was now nearly bare but for shredded wood and a few oases of green I'd asked Adams to spare; I felt a little queasy at the idea that shortly before, the hillside had been mobbed with vegetation. I walked in rows, looking for mangled stumps, but there were none. There was no trace, no root ball, no evidence the autumn olives had ever existed there. I knew it was unlikely that all the roots had been dug out. I expected to see new growth in spring. As the weeks passed, I looked for it. This slope was now the definition of a disturbed area, and if any invasives were in the seed bank or nearby, which I assumed they were, I could easily end up with a hillside covered with weeds. This time, I told myself, I'd stay on top of it. I'd pull any seedlings before they could grow large—now I'd be able to see them and respond quickly.

When I saw that a small grove of young elderberry bushes I'd flagged for Adams to spare was still standing, I celebrated; elderberries were my favorite plants on that slope and among the most valuable for native wildlife. The elderberry, or *Sambucus canadensis*, provides food and habitat for forty different species of caterpillar. The berries, which emerge in July and August, are favored by all sorts of creatures, including wild turkey, bears, and box turtles. Among birds, the indigo bunting, red-eyed vireo, red-bellied woodpecker, and eastern towhee, all of which live on the mountain, are fans of the elderberry bush.

Every week, I checked on the elderberries. I worried that they were too exposed to deer, now that much of the surrounding vegetation, including protective thorny bramble, had been cut away. On subsequent visits, I found more of the bushes coming up, until there were at least a dozen in various stages of growth. *See*, I thought, *the elderberries are protecting themselves from hungry deer by multiplying*. Which was a bit of magical thinking, probably. I was feeling attached to these plants because they were something that worked; they were actually supposed to be here, and the fact that they were thriving felt like a small victory. In the coming weeks, I watched them flower. The tiny white buds gather into an upside-down saucer; until you come close, each "saucer" appears to be a single large flower. The flowers smell delicately sweet, not pungent like the elderberry syrup I'd been advised to use as a home remedy when my kids had colds. When the bushes were in full flower, I could stand high above at the top of the slope, a couple of acres away, and still see the white blooms.

Crownbeard, milkweed, and blackberry grew to fill in the rest of the bare slope, a welcome sight. Maybe I shouldn't have been surprised, since these plants grow in abundance all over the meadow. The real surprise was: no weeds, to speak of. I began to wonder if wiping the slate clean would be worth a try in other parts of the meadow that were beset with invasives. Week after week, I watched milkweeds grow and flower in clusters around the elderberries. There was yarrow, and there were tiny asters that had trouble pushing through the layers of dead grass in some of the upper fields.

This steep slope had become a prime example of what I'd hoped for, and what some of my experts had hypothesized—a high concentration of native plants had waited in the soil for an opening, a chance to emerge and shine.

"What's there first matters more than what you do," said Nancy Edwards, a grassland restoration specialist who led prescribed burns, managed deer hunting, and handled occasional bush-hogging on a large research property. Nancy (a pseudonym) wasn't advising me to give up on trying to influence the results, to give in to fate; she was reminding me that the plants I see when I walk the field, plus those in the seed bank, will, in large part, determine my course of action and predict the outcomes, no matter what conditions I might aim for. So far, most of the native plants I was seeing on the hillside were sturdy, hardy ones that could grow and compete successfully alongside nonnative fescue. I saw fewer of the more fragile, sensitive native plants that are pickier about where they grow.

I kept looking for signs that the autumn olives were returning. I didn't find any. I was certain I'd spot them before they became too large to pull with my trusty weed wrench, which I nevertheless couldn't imagine lugging down, and back up, that slope.

As the weeks went on, midsummer came and went, and I relaxed my vigilance to manage other fields where I'd noticed unexpected changes occurring: where crownbeard had ruled, there was an explosion of dogbane. Was this good? I wasn't sure. I found signs that poison hemlock—an invasive weed, and, yes, the plant that was used to kill Socrates—had spread to a new part of the meadow along a mowed path. Definitely bad. On the plus side, a profusion of flowering wild bergamot (*Monarda fistulosa*) on one hillside in July led to days of bumble bee watching. In this same field, I spotted my first bumblebee moth, a creature that from a distance resembles a hummingbird with a bumble-bee-patterned black and yellow body, but turns out to be a moth that drinks nectar through a long proboscis while hovering beside a flower. On the down side, I lamented what seemed to be a drop in the number of milkweed patches. I searched, but failed to find many of those I'd seen the year before. I didn't learn until later that succession was probably part of the cause; milkweed and crownbeard were less abundant in some of the fields where shrubbier plants like dogbane and blackberry had begun to take over.

One day in late August, I stopped at a plateau near the old oak, about fifty feet before the steep slope where Adams had mowed. That spot had always been filled with mixed growth, some good, some bad, in part because it bordered an area that was regularly disturbed by mowing; any seeds ejected by the mower would blow right into it. There was crownbeard and dogbane, and a few pokeweed bushes, intermingled with lots of nonnative thistle, and the only weed I'd given in and sprayed directly with herbicide, that poison hemlock. This was not a pristine patch of ground. Even so, I was disturbed by what I found that afternoon: a skinny vine I'd never seen before climbed over the plants, winding up the thick, square stem of the crownbeard, over and around the pokeweed bushes, and crawled over anything that was closer to the ground. The vine didn't discriminate—native and nonnative plants alike were victims. The taller plants sagged and bent under the vine's weight. It almost seemed as if one endlessly long strand had knitted row after row, climbing over itself again and again to form a wide blanket, smothering the plants underneath.

The vine's leaves were triangular, like party hats a child might draw, and an innocuous Kermit-the-Frog pale green. They were linked by stems that were sometimes red and sometimes green. Tiny disks made of leaf material encircled the stems at intervals along the vines. The stems were covered with what appeared to be tiny hairs, but were actually barbs—not thorny like blackberry, but grabby like Velcro, and they clung to everything they touched: a long strand stuck on my sleeve and attached

Unintended Consequences Farm

itself to my trousers. The vines stuck to the plants they'd climbed up and over and around; they even stuck to themselves, forming a dense web, if a web were not delicate but instead a mat of barbed silk, impenetrable. The vines began everywhere and nowhere, a sticky net tossed over the plant life on that plateau. When I tried to pull strands of vine off the plants with my gloved hands, it was like peeling the plastic seal off a microwaved dinner, if the plastic had melted and stuck to the food (wait, maybe I was supposed to remove the plastic first . . .). Some of the plants underneath came with it; most didn't. But the plants that were left on the ground looked clobbered, like an ATV had run them over. I couldn't tell where the vine started or ended—it kept coming and coming. It covered a larger area than I'd first realized, at least fifty feet deep. When I found places where it grew out of the ground, it wasn't hard to pull it up by the roots, but it was hard to find those places amid all the vegetation. If I couldn't remove the vine, it would kill the plants underneath it, suffocating them and blocking their light. I didn't have any tools with me, but I was driven—I kept pulling the mats of vine until I ran out of daylight. There are days when I doubt my preparation for this task. That was one of those days.

I had been watching for the wrong plant all along. This was mile-a-minute, *Persicaria perfoliata*, an invasive vine that has made it onto noxious weed lists in seven states so far. Some common plant names are baffling to me or simply dull, but this plant's name is descriptive, though it doesn't grow a mile a minute; it only seems that way. But bad enough—it grows six inches each day, pulling itself up and over vegetation, and it can reach as high as twenty-five feet, climbing up trees at a forest's edge. It's an annual, so the existing plant wouldn't return, but it would die in place and smother everything beneath it. And before it died, it would drop enough seeds to guarantee its forceful return the following spring—and the spring after that, and the spring after that, and on and on.

The vine originated in East Asia, and it's so vigorous I'm surprised it didn't just stretch across the ocean from Beijing in one long strand.

At first, it was intentionally introduced as an experiment in Oregon in the late 1800s, and, again experimentally, in Maryland about forty years later. I wonder what they were trying to learn from those experiments and what they planned to do if their efforts were successful. But I wasn't able to find out. And the experiments didn't work—not until the plant was introduced in the 1930s, this time by mistake, in contaminated holly seed or imported rhododendrons. Now, it can be found from New Hampshire to North Carolina, and all the way west in Oregon.

In its native China, mile-a-minute vine is not as much of a pest; it's kept in check by insect predators. But here, there's nothing to bug it save for a tiny weevil that bores through its stems and eats holes in its leaves. The weevil might slow the vine's roll, but it won't kill it off entirely. Only humans can do that. Or, they can try. Pull it, spray it; it will keep coming back until its seed supply is exhausted. Even then, it would return, its seeds carried by birds and deer and muskrats and tractors and transported long distances by streams—its seeds can float in water for up to nine days.

I'd need to act quickly if I didn't want the vine to spread beyond the plateau and stretch itself down the slope, potentially shading out and killing the native plants that had grown up where Adams had cut out those autumn olives months before. I returned a few days later with sturdier lawn bags and my heavy rake. I raked and bagged the stuff until my clothes were soaked through with sweat—good thing my shirt was made of that fabric that claims to wick moisture away from your skin; I only felt like I'd just climbed out of a swimming pool wearing a wet poncho. I was exhausted, and I'd hardly made a dent. It was summer; the vine would continue to grow until the first frost. Native pokeweed berries were more nutritious for the birds, but birds couldn't access those berries amid the layers of vine that were drowning the plants. Or maybe I was the one who was drowning. I needed help.

I met with a landscaper who said he could get rid of the rest of the vine before it spread, no problem. We stood on the plateau, and I asked him how long he thought the job would take. He didn't answer right away; instead, he walked, slowly at first, then haltingly down the slope, past the field that had been cleared. He stopped and looked right and left. He muttered something and swiped the hair away from his forehead. He kept walking. I followed down the hill, and then I saw what he saw. The vine already wound down that slope, through the fields where the milkweed and crownbeard had been growing energetically, and it continued even beyond that. We followed its progress, down, down, until it became clear that the vine didn't stop until it reached the shade of the woods all the way at the bottom, acres downhill from the plateau where I'd initially spotted it. It had already taken over the hillside. How had I missed it? I'm guessing it made its entrance from isolated spots along the field's edge, or from farther down the slope, below the cleared field. At a distance, it had been camouflaged by its delicate appearance amid all the shades of green on the slope, but up close, shrouding everything in a dense layer of light green, it was undeniably apparent. While I was busy watching bees and butterflies and bumblebee moths, wandering amid wild bergamot clear at the other end of the mountain, a vine ate the hillside.

I've gone from saying I'll get rid of the invasive plants on the mountain to saying I'm *managing* them, as if they're bankable but unruly child stars. I've come to face a difficult truth: that to aim for completely eliminating these plants is a recipe for disappointment and defeat. I don't want to pitch out idealism—wishful thinking is part of what drives me. But if I don't want to lose my mind, I need to work instead toward reducing the plants' numbers and keeping them from establishing in new locations. I *had* come to terms with this, I *had* settled on a level of acceptance, if not quite peace: I'll still feel the cortisol surge when I find spotted knapweed in a new spot, but, I tell myself, I can't do everything, and I can't be everywhere. What I can do is try to approach this work in a way that seems logical, by focusing first on the areas where I'm most likely to be successful in making a difference. What I mean is, I'd begun to come to terms with imperfection. This is hard for me.

I attended a talk on the importance of native plants, presented by a regional conservation organization, in which the featured speaker noted that he removed *all* of the invasives on his ten-acre property. I was surprised to hear that, and I felt certain that he wasn't being literal. I was wrong. During the Q&A, I asked what the organization saw as a reasonable balance of natives and invasives to aim for on large acreage, given that elimination was probably not realistic. I said I'd heard that 80:20 (80 percent native plants) was a good target, and even that felt like a fantasy to me at that moment.

Douglas Tallamy advises aiming for at least 70 percent native plants, due in part to the documented impact on birds if the percentage falls much lower. Charlotte Lorick, the plant specialist, says that even 50:50 can be fine, depending on what you've got. Say, for instance, in your suburban yard you have a patch of white clover instead of lawn. White clover isn't native, but it's better than fescue. Generalist bees will gather pollen from white clover.

On the other hand, 25 percent of the bee population are specialists—bees that will only go to a few, and sometimes only one, flower species. A recent survey counted 183 specialist bee species in twenty eastern states and found that North Carolina, Virginia, and Maryland had the highest numbers of those species. Those bees are in decline, and they won't touch white clover. Nancy Edwards told me that on the several hundred acres of land where she worked, they were in the process of eliminating clover in favor of native plants. So, whether 50:50 is good or bad, or 70:30 or 80:20—like so much else, *it depends*.

"In my experience, we can do better than that." That was the answer a representative of the organization gave me during that native plant talk's Q&A. I'd thought 80:20 would be seen as a gold standard; and I was only aiming for 75:25, based on conditions on the mountain. The rest of his answer surprised me even more: he said it was reasonable to expect to eliminate *100 percent* of invasive trees, shrubs, *and vines*. He admitted that annual forbs like spotted knapweed, and grasses, like stiltgrass, can take a long time because they last for years in the seed bank, sometimes five years or more. I could see nearly eliminating trees like ailanthus and paulownia, and maybe autumn olive, first of all because I could find most of them, and because there are ways to treat them without harming nearby plants.

But vines? I thought of each field I was supposed to be tending, along with the forests whose edges were rife with Asiatic bittersweet and Japanese honeysuckle, both of which are invasive vines. I brooded grimly over that mile-a-minute suffocating the hillside. What could I do but laugh, a sort of choking sound. I felt like the narrator in a short story I'd read by Lorrie Moore who, in response to a situation that she finds not funny-funny but funny-sad, does nothing but laugh for more than two pages: "Ha! Ha! Ha! Ha! Ha! Ha! Ha! Ha! Ha!" I think I could laugh wryly for an entire chapter about vines.

At nearly every turn, it seems like the one small bit of good I think I've done is thwarted, overshadowed by the harm of what I misjudged or, in this case, just plain missed. Brian Morse tells me there isn't a single property in the county where he hasn't spotted mile-a-minute. The seeds are too easy to spread. The best I can hope for now is to keep it from making its way to other parts of the meadow. How to do that? Avoid the areas where the plants are releasing seed, and clean off clothing, trucks, and equipment before going to another part of the property. Easier said than done.

On the hillside that was cut, a central path down the slope is consistently maintained to allow me to access the lower fields and forest, although it's so steep in places, walking that slope is a real challenge. If every property in the county is infested with this vine, any other property that's been mowed was probably infested, too, and I suppose seed could have been introduced that way at any time in the past. But it's more likely that mile-a-minute was already here in the seed bank somewhere, but I didn't know it, and forestry-mulching the hillside—a new disturbance that left bare ground and created light in places that had been shaded— allowed for a "plant explosion." In the end, I'd never know how it happened, and it didn't matter. I only know that the year before, I didn't see mile-a-minute on the hillside. Maybe there was too little of it for me to notice. Maybe it was farther downslope where I seldom ventured. Most likely, I wouldn't have known to look for it at all.

In a streaming series narrated by David Attenborough, *The Green Planet*, there's an episode in which botanists hover in helicopters over thick tropical forests on the island of Kauai in Hawaii and target invasive plants from the air. From the helicopter, a sniper shoots a high-precision paintball gun loaded with herbicide balls instead of paint. This is an ingenious (and extremely expensive) way to carefully target invasive plants in a place that's otherwise too difficult and impractical to access on foot or by vehicle.

Maybe a paintball gun loaded with herbicide is exactly what I need, though I'll have to do without the helicopter. If the effort is narrated by David Attenborough, I'll try anything.

Mile-a-minute is listed by the U.S. government and the European and Mediterranean Plant Protection Organization as a noxious weed. The State of Virginia recently added the vine to its noxious weed list, which is administered through its Department of Agriculture and Consumer Services (VDACS). That list is a separate entity from the DCR's invasive plant list, and it's surprisingly short and notable for what it doesn't include—autumn olive, for one. At first read, the VDACS list makes it sound like selling or planting listed plants could be grounds for a misdemeanor charge. In reality, it's reversed—the law prevents a plant from being listed if it's currently sold by a nursery somewhere in the state. The autumn olive was, until recently, one of these plants. It was still sold by a nursery, so it couldn't be considered for the list. Now that it's no longer sold in the state, it's been approved to add to the list, but even once a plant is approved, the process can take time. (Note that a plant can't make it onto the Maryland list *unless* a nursery is selling it, and it can't make it onto the Virginia list if a nursery *is* selling it. Are you confused yet?)

There are three possible tiers under which a plant can be categorized on Virginia's list. Tier 1 is for plants that would pose a threat, but aren't yet found in Virginia. The goal is to keep them out of the state. Tier 2 contains plants that are found there, but not widely, and for which there's a reasonable approach to control—suppression or eradication. Mile-a-minute vine is classified as a Tier 3 plant. Tier 3 is for plants whose spread may be slowed by restrictions on their movement. I'm trying to imagine how such restrictions would be implemented, when you can spread a plant's seeds simply by brushing against it. (Not surprisingly, there's a form to fill out.) But the part of the Tier 3 designation that stops me is this: it includes plants for which eradication and suppression are "not feasible." In other words: *abandon all hope.*

In better news, in 2023, two new laws were passed in Virginia that propose promising changes to the way the state handles its noxious weed lists and implementation. First of all, the legislation appears to give the DCR list some teeth: the list will be updated every four years—the current list hasn't been updated since 2014—and no state agency can plant or sell anything on the list. The new law restricts use of listed plants by state agencies to scientific or educational use, and it gives VDACS authority over permitting for moving or selling the plants in specific situations. And, perhaps the most important development, any tradespeople, like landscapers and landscape architects, who purchase plants for use on private land must inform the landowner if those plants are listed as noxious weeds before installing them. Two-thirds of plants sold to landowners are sold through tradespeople rather than directly from retailers, so this law should have a significant impact both in educating consumers and reducing the invasive plants out there. The law stops short of requiring retailers like nurseries and big box stores that sell directly to consumers to let them know when they're buying an invasive plant, and it doesn't prohibit the sale of these plants. In 2024, the legislature passed a bill that would have required all retailers to post signage informing customers about plants that are invasive; unfortunately, the governor vetoed it. Until such a law is enacted, it's important for consumers to ask questions and do a little light research to make sure they're not buying plants that, while they may be legally sold by nurseries, pose a serious threat to native ecosystems.

The second new law mandates that state agencies work together to plant native plants on state lands, including restoring selected state lands with native plants. A similar law is under consideration in the U.S. Congress, recommending the use of native plants on federal lands. It may be surprising that such laws are not already on the books, since it seems "natural" for governments to set an example when it comes to native plants and habitat restoration, but growing popular awareness of the importance of native plants for a healthy ecosystem and for food pollination is a relatively recent development, and the idea that a landholder as large as a government should respond and play a role stems in part from that awareness. So here's hoping. New laws like this could effectively prevent another autumn olive or kudzu debacle, and I can't help thinking that if my hillside had not been threatened by autumn olive to begin with, I might not now be confronted with a rampant invasive vine.

However the mile-a-minute infestation had started, the infested area could no longer be described in square feet; it had to be estimated in acres. Worse yet, I soon learned that the vine was already in seed. Any further disturbance now would be inadvisable because it would spread the seeds. The more advice I sought, the more it was drilled home to me that there was nothing I could do. I had to wait for the right season to come around when it would be safe to act. Even when the plant has died off for the season, the seeds remain viable. Dead, the vine stays there, draped over plants like a shroud and preventing new plants from growing by keeping them in the dark. I was back at square one, or more like square minus-ten. It might have been better if I'd never done anything to that hillside. I'd still have a bevy of giant autumn olives marching toward the upper meadow, but even they were easier to deal with than this vine. I stared down the mountainside at the pale green blanket that seemed to cover everything. I couldn't find my cherished elderberries anywhere.

Chapter 9

OH, DEER

... in which weeds follow the herd

I'm not good at waiting; I need to feel like I'm doing something, so I pull invasive Japanese stiltgrass instead. Late August is the right time for that, at least, before it starts to put out seed. And stiltgrass is easy to pull, so at first it feels rewarding; the roots are weak and the whole plant comes away with a sweep of the hand. Which is good, because there's a lot of it. Under the old cherry trees, where the ground is moist and partly shaded, there's more than there was the year before, even though I pulled it then, too. That's because, like many invasive annuals, its seeds remain in the seed bank for years. I'm also seeing stiltgrass appear for the first time along a steep hillside I call the "sledding hill." A long skinny field sandwiched between woods and a path that's mowed regularly, the sledding hill is clear on the opposite end of the farm, more than a half mile from the mile-a-minute-infested hillside.

Stiltgrass leaves sprout along a central stalk, and what appears to be a white stripe runs up the center of each leaf; this stripe helps to differentiate it from other grasses, but its shade of green is close enough to that of intermingled goldenrod and wild bergamot that when I pull, I have to take my time and carefully separate the weed from the native plants. The stiltgrass grew up from underneath these nearby flowers and then matched and surpassed them in height and abundance. It shades out, crowds out, or stunts its neighbors. This is another reason why spraying stiltgrass would be a tricky operation; beneficial plants would undoubtedly also be sacrificed. Where the grass is over a foot tall, I can't see what's underneath it until I've pulled it, and more than once I stick my hand into a hidden bed of poison ivy.

On the sledding hill, I'm amazed to find stiltgrass that's over three feet tall. I begin to clear it away, and in the process I uncover a pokeweed bush that had been entirely obscured from view. Pokeweed can grow up to six feet tall, and its berries are a valued, nutritious food for fox and turkey and everything in between. Another native species that likes to hide in the grass leads me to prod the ground cover with a shovel before I reach my hands in. This time of year, copperheads coil amid leaf litter and in the shade of tall grass. A neighbor tells me he surprised two of the snakes while pulling his stiltgrass; I proceed with caution.

Japanese stiltgrass (*Microstegium vimineum*) arrived here as packing material in crates of porcelain shipped from China (not Japan) in 1919; it's also said to have been used as grassy packaging for bottles of imported wine. I imagine people unpacking the crates dockside and tossing the plant matter to the ground, where it dropped seed and grew. The plant was first identified in Knoxville, Tennessee, and made it to Virginia by 1931. Still, it took decades to become a noticeable problem in forests. The theory is that as human disturbance ramped up, the more easily invasive plants like stiltgrass could make inroads into native plant territory. It's now listed as a noxious weed in more than forty states; it does not yet appear on the VDACS official tiered list, although it would seem like a shoo-in for Tier 3.

An invasive plant's superpower is its adaptability. Some improve their chances of success through multiple self-propagation methods, and Japanese stiltgrass is one of these plants. Seeds that form at the top of a blade of stiltgrass are spread by wind and released by the slightest disturbance.

Oh, Deer

A deer or a dog brushing against it will transport the seed on its fur; water washes the seed down your ditch and into your neighbor's woods, or into a stream where it plants itself alongside the streambed. And then there are the leaf blowers.

You would think that mowing shortly before the plants go to seed would prevent them from spreading, but the plant has a fix for this: it produces seed in a second spot, closer to the ground. No matter how low you set your mower, the plant will seed lower until it's practically seeding at dirt level. The only way to prevent this is to use a weed whacker and aim it so the string is hitting the dirt as you cut the grass. The bare spots left by weeding or whacking could become openings for other invasives, or they could allow native flowers to grow and spread instead.

When I weeded the stiltgrass away from that pokeweed bush, any of the grass I touched but hadn't yet pulled immediately fainted—flopped to the ground as if I'd sat on it, making it more difficult to avoid pulling the wrong plant. This response enables its third method of reproduction: stiltgrass can sprout new plants from the nodes along its stem wherever they touch the ground.

It's hard to demonize a plant that's designed so intelligently. In their native habitats, most of these plants have to survive threats from multiple predator insects or fungi, and they've evolved these multiple methods to compensate; if one doesn't work, they still have the other one—or three. At home, they're subject to predation and disease, and their abundance is more likely to be controlled, but even in its native China, where stiltgrass is largely innocuous, it can become a pest in agricultural fields. A disease-carrying fungus that scientists hoped might prove to be an Achilles' heel for stiltgrass in the United States was unfortunately found to attack native species as well; instead of hurting it, it can actually help stiltgrass spread. So much for that.

There are no inherently bad plants, only plants that are in the wrong place. Somewhere on Earth, plants that become invasive here are largely harmless native plants, not carnivorous monsters from outer space that feed on secondary characters in a Broadway musical. In a plant's native territory, there are local wildlife that use it for food and shelter, and few battles are being fought or chemicals patented or biocontrols released in order to hasten its demise.

It's easy to forget this when we see some of the hyperbolic language used to discuss invasive plants. Language that frames a plant's behavior as willful, and the damage it does to the environment as intentional, may be catchy but it can be harmful in itself. In their native habitats these plants are usually part of a balanced ecosystem; where they aren't native they turn scourge, run rampant, consume every square inch of land and eventually conquer Rome ... Whoever can talk about these aggressive introduced plants without using any negative language might deserve a medal. Even the word "invasive" has negative connotations. Sometimes the plants are instead referred to as "introduced," which is accurate but incomplete. Since not all introduced plants become a problem, that term is inadequate without adding a modifier like "aggressive," as I've just done, but that, too, is a negative. Perhaps "too successful" would be acceptable, then? When you start parsing the language, it's easy to get lost in the weeds. That's why, when it comes to talking about invasive plants, I prefer the catharsis of simple, straightforward profanity.

The plants haven't consciously chosen a destructive path—humans have chosen it for them, reinforcing it with our behavior over decades and centuries. Our own native plants can become pests, too, in the wrong environment. Broomsedge was introduced in Australia and Hawaii and is now invasive in those places. More than twenty years ago, a fire known as the Broomsedge Fire burned more than one thousand acres in Hawaii Volcanoes National Park on the island of Hawaii. Unlike native grasses here, Hawaii's native plants are not fire adapted, but since fire-adapted grasses were introduced to the islands for cover crops and forage, they've not only led to events like the devastating fires on Maui in 2023, they're the first to grow back and take over after a fire, setting up the next dangerous fire event. These plants now cover 25 percent of the Hawaiian Islands. Those who frame invasive plants as the villains have fingered the wrong culprit.

Oh, Deer

It's not as if I'm starting a Misunderstood Stiltgrass fan club, and I'm sure that I occasionally fall into exaggerated language myself. If descriptions of invasive plant mitigation sound like tales of battles and wars, it's because that can be an apt metaphor. These plants displace and inhibit the growth of native plants to the detriment of everything that relies on those ecosystems including, eventually, us.

Where there's already compact, crowded plant growth, stiltgrass is usually kept at bay. The places where it grows on the mountain have something in common: they adjoin areas that are regularly kept clear and see frequent traffic (people, deer, trucks), or they're near drainage areas, like culvert outlets. The plant could easily move into the woods from these spots, and that's what I want to prevent. Stiltgrass usually prefers shade and moist ground, and the woods offer both. Oddly, though, I'm also seeing the weed in bright, sunny places. Celia Vuocolo mentioned that it's unusual to find stiltgrass in these sunny locations, but for unknown reasons, it was an unusual year; she'd noticed stiltgrass luxuriating at the edges of sunny fields all over the state. What if it's adding yet another adaptation to its many offensive moves? Now I have to worry that it could spring up throughout the meadows, too.

Stiltgrass can easily take over a forest understory, smother native plants, and form a monoculture. The classic infestation resembles a green sea filling the forest floor. But isn't something growing better than nothing? Hungry deer have hoovered up the native understory and middle story in these woods—the ground cover as well as the small trees and shrubs that grow in the shade. Unfortunately, the deer will bypass a field of stiltgrass in favor of native oak seedlings any day. Why not let something, anything, grow there, and study the new ecosystem that forms around it? Isn't it natural for plants to compete and for some plants to win?

Similar to the autumn olive's nitrogen-fixing ability, stiltgrass changes the soil chemistry where it grows, repelling beneficial insects and encouraging undesirable mites. As stiltgrass takes over the forest floor, the soil composition and balance are altered to favor the invasive plant, increasing the odds that it will face little competition.

Is this a case of "all's fair"? No. Where plant variety is eliminated, wildlife will become scarce because there's nothing for them to eat and no place for them to shelter. According to the National Audubon Society, invasive species are the biggest threat to birds here in the northern Blue Ridge. Plants like stiltgrass, and nonnative plants in general, threaten biodiversity: wildlife will normally avoid plants they didn't co-evolve with over thousands, or millions, of years. A single species of native plant may support hundreds of helpful insect species, whereas stiltgrass supports *zero*. This may explain why, amid all the stiltgrass I pulled over days and weeks, I didn't encounter another living thing—no bees, no beetles, no spiders, not even a fly.

***

Shortly before Matt hiked through the forest here, two foresters from the state Department of Forestry examined the condition of the woods and advised me generally on how to improve the native understory. They hoped to see mosses and ferns and a gazillion seedlings from oak, hickory, and birch in those woods. But they didn't; these plants are scant or absent in many parts of the forest, and invasive plants from the adjoining meadows and elsewhere are beginning to make inroads where the forest floor is bare.

Following their visit, the foresters gave me a one-page form containing three recommendations to be implemented over the next few years: The first year, I should control invasive plants in and around the forest (ha! ha! ha!). The second year, once invasives are taken care of (oh, the optimism!), I could start encouraging the growth of a native understory and middle story. But the DOF's third directive could not have been more blunt—before I try either of these steps, they wrote: "Harvest deer. As soon as possible."

There are many, many deer on the mountain. I'm not sure how many, because in the country, unlike the suburbs, they can afford to stay away

Oh, Deer

from me, and they aren't good about standing still so I can count them. A Smithsonian Conservation Biology Institute (SCBI) study of long-term deer-browsing impact in a nearby forest estimated deer density in my area at a fairly consistent forty-five to sixty deer per mile over the past thirty years. That's three to four times the number of deer that are usually considered desirable. Northern Virginia, which is more densely populated with humans than this county, is host to even more deer than that: sixty to one hundred per square mile. To put it in perspective, five to fifteen deer per mile is the optimal number I've most often seen mentioned; the difference in the desired number depends on the place and may have as much to do with deer impact on the human population as their impact on native plants or the availability of food.

There are two ways to measure whether you have too many deer. One is cultural carrying capacity (CCC), and the other is biological carrying capacity (BCC). BCC relates to the food supply—how much food there is in the area to support deer—in other words, are there plenty of hostas in your yard for them to eat? CCC describes how well, or not, people in a given location will tolerate the abundance of deer, in other words, are they eating all of your hostas, and how do you feel about that? The BCC, the actual availability of food, is almost always higher than the CCC. Some people enjoy seeing deer wander down their streets, but they don't enjoy providing their gardens for lunch or seeing deer carcasses along the road. They especially don't enjoy Lyme disease–carrying ticks, which acquire the offending parasite from mice and are transported by deer like a giant rideshare. On the mountain, the relevant measure is BCC; in the woods here, they're eating the local wildlife out of house and home, literally.

Deer will eat everything they like within their reach, or up to five feet high; this is known as *browse height*. In forests where the middle and understory have been overwhelmed by deer, you can usually see a browse line at around the five-foot mark. Everything above is intact and leafy; everything below is munched, or missing, with the exception of most invasive plants. Since the deer haven't evolved to eat them (and many are none too tasty), those will be left alone. Deer will occasionally eat foods outside of their normal diet, but generally not unless other food sources are depleted. For instance, in a forest where they've already eaten the herbs and the oak seedlings and the maples, deer may begin to nibble on beeches, which they otherwise avoid. They'll sometimes snack on invasive kudzu, though not enough to make a dent. But deer will never eat stiltgrass or garlic mustard, two of the worst culprits that form monocultures on the forest floor; there are no native creatures here that eat either of those plants. Garlic mustard was initially introduced as a food plant for people, and even though deer might sometimes browse our garden plants, they steer clear of the stinky ones, so garlic mustard is out. (Oddly, and sadly, garlic mustard attracts a rare butterfly in West Virginia that is enticed to lay its eggs on the plant by mistake. The caterpillars hatch and they can't tolerate the chemicals in the plant's leaves. They can't eat or grow, and they die.) Unfortunately, what do I have along the edges of the woods in multiple locations? Stiltgrass and garlic mustard.

The dual impact of deer and stiltgrass drastically altered the middle- and understory in a forest, in a study of the interaction between the two led by community ecologist Benjamin Baiser. Together they rendered the forest floor an unfriendly place for native bird species that nest at or below the browse line; as a direct result fewer of these birds were found in infested forests, and they declined in greater numbers than those that relied on the upper canopy for habitat. The researchers worried that these effects on the forest could be long-term and prevent forests from regenerating. Meaning, seedlings wouldn't be around long enough to grow into saplings or mature trees, and mature trees that died wouldn't be replaced by new trees of the same species. In a separate study of islands in British Columbia over a fifty-year period, songbirds were 55 to 70 percent less numerous on the islands that were browsed by deer. And we already know that many forests are essentially islands surrounded by developed land.

Sometimes even nine deer per square mile is too many. At that level, according to the Massachusetts Audubon Society, the number of wildflowers begins to drop, allowing invasive plants to take hold. When the flowers dwindle, deer eat plants with tougher leaves and woody stems, again leading to habitat loss for birds. Mass Audubon estimates that as many as a third of migratory forest birds could disappear as a result of the high density of deer. Two birds that may be affected, the white-eyed vireo and the prairie warbler, spend spring breeding seasons on the mountain and need forest understory to complete their life cycles. The warbler is already considered a species of concern because of a reduction in habitat.

All of this dining above ground has a notable impact on what's happening below ground, too. When everything below the browse line is eaten, there are no plants to drop seed, and that means fewer seeds from native plants stored in the seed bank. Scientists at Cornell University found that deer browsing caused a 5 percent decrease in native plant species in the seed bank when compared with the incidence of nonnative plants, and a 12 percent decrease in the abundance of native plants. Deer not only change the field and forest in the present, they can predict its future, too. Chalk up another win for invasive plants, another advantage they don't need.

This mountain is defined by what deer do and don't eat. There's blackberry, which is native, and multiflora rose, which isn't. Both are thorny and form impenetrable bramble, and if I want to walk off the beaten path, I need to carry loppers with me to break through the thicket. Unfortunately, deer aren't interested in dining on thorny plants, only eating the berries, which spreads the seeds.

Most of the time, though, deer don't waste much—they'll eat the leaves, flowers, buds, and fruits. The mountain is full of deer favorites: honey locust pods, acorns, persimmons, hickory nuts, flowering dogwood, and oak, birch, and poplar. Deer require as much as six pounds of food each day for every 100 pounds of body weight. If a doe weighs 150 pounds, she'll eat at least 9 pounds of food a day. Now multiply that by ... how many deer?

Here, I'll have to cage oak seedlings and other plants until they grow above the danger height if I want to save them. The white oak is one of the deer's favorite woody meals; its sweet acorns are their favorite protein food. Some forests here that used to be dominated by oak are now being overtaken by maple, which the deer don't seem to care for as much. This may not seem like a big deal—most maples are still native trees—but a maple-dominated forest is a very different place than an oak-dominated forest. Oaks are keystone species, crucial to the food web, in particular for nesting birds. If deer continue to eat most of the seedlings, the oaks won't replenish, the forest will thin out as older trees die, and fewer ecosystem-important trees will take over. Oaks grow slowly; every seedling that doesn't make it to five feet is a potential years-long setback. The future of this forest and the wildlife that depend on it will be determined by which tree species are best at surviving these threats.

In a world made up of islands, deer are in their element: edges are a deer specialty. The development of suburbs and agricultural fields and office parks created innumerable edges for them to browse. Here on the mountain, while they browse along a streamside and a mowed path for ragweed, pokeweed, and wild strawberry, they deposit stiltgrass seeds, too.

The SCBI found that when deer were excluded from an area of forest with fencing, the volume of stiltgrass growth inside the fenced area was dramatically reduced, compared with areas where deer were not fenced out. So, fewer deer does seem to mean fewer invasive plants, if not fewer ticks. (Reducing the number of deer unfortunately only increases the number of ticks hitching a ride on each deer.)

One local farmer reported watching a doe teach its fawns to break open a watermelon with their hooves. The farmer ended up erecting a mile-long deer fence around her farm to save her crop. It's hardly practical for me to put an eight-foot fence all around the mountaintop. Even if I could do it, fencing might exclude other wildlife, too, like bears; although, unless the fence is electrified, it probably won't stop a bear from going over the mountain when it wants to.

Some of our adjoiners allow hunting on their land. I wonder if the deer have figured out that they can come to our meadow to avoid being hunted and find plenty of food and shelter here. Research shows that deer are more than twice as afraid of human voices as they are of most other predators. Wolves inspire only the second highest rate of fear; this was true even

Oh, Deer

in cases where there were no wolves living in the region, which implies that it's instinctual rather than learned. But deer are more worried about us. Hunters have sometimes blamed coyotes for scaring off deer, but based on the research, the blame may fall on the hunters themselves.

Most everyone advised me to invite deer hunters up here to help reduce the herd, and I'm now resigned to that necessity. But before I take up hunting myself, as some have suggested, I need to get past my longtime bias in favor of the deer. How is it a fair contest when the animal has no defense against a firearm besides an ability to run? What sort of person, I used to think, would shoot an animal with glistening soulful eyes and a narrow muzzle and shiny nose that remind me of my dog? I know it's not that simple.

***

In the suburban Maryland house where I grew up, our kitchen table was stationed in front of a set of sliding glass doors that led to a patio. From the table, we could gaze through the glass at our backyard lawn and trees, and just beyond, directly into the yard of the neighbors whose house backed up to ours, with their similar patio, their charcoal grill and their lounge chairs, and their similar glass doors. It was a neighborhood with a homeowner's association that discouraged tall fences, and at that time our fence was picket height, so it offered little privacy, but we liked the illusion that our yard was larger than it was; there was nothing to stop the eye until that house behind us.

Our backyard neighbor, whom I'll call Charlie, had built a dog run to house his three pointers, along with any new litters of puppies. He was a deer hunter. In season, he'd load his pointers onto his truck and take off for the country. (For a few years, his hunting vehicle was an old hearse. He parked it in his backyard, in view of my grandmother's bedroom window, which led her to lament that "they" were coming for her.)

One fall evening while it was still light, we sat down to dinner in the kitchen and gazed absently out at the yard like always. Charlie must have returned from one of his hunting trips, and before we realized what was happening, he had strung up a deer carcass between two trees that lined up precisely with our view. The sight was as clear as if it had been a few feet away on our own patio. He proceeded to skin the deer. I saw the carcass shudder as he slid the knife under the skin, the skin folding away from the body. My brother said, "What the heck does he think he's doing?" (He never cursed.) My grandmother said, "Oh, he has a deer there," in the same tone she'd use to say, "There's some lipstick on your cheek." We were all paralyzed for a minute. Then my mother wordlessly stood up from the table and drew the curtains, as if it was only the light in our eyes. This incident may have influenced my feelings about hunting.

Here on the mountain, a family of deer browses on a south-facing hillside. One of them, far off, vigilant, turns its head in my direction. Another stands at close range, alone under the apple tree; it sizes me up and ambles away. I don't have to see the deer to know they've been here. There are the depressions in the grass where a doe has hidden her fawns; the shredded bark of a birch—a buck rub, where a buck has left a message for other deer; the pellet droppings in the field, of course. And most of all I witness it in the depleted forest understory, where, in some parts of the woods, leaves and rock crunch beneath my feet and not much else. The overabundance of deer has changed the face of the land.

When the English arrived in Virginia in the early 1600s, they arrived at a place that was inhabited by Indigenous people who may have lived on the land for as long as fifteen thousand years, and who didn't believe in land ownership or possessions the way the English did. Native people didn't view the deer and other animals they hunted as "owned" by anyone. This must have seemed like a great advantage to the colonists; as far as they were concerned, they owned the land—it came from the Crown. And if the native people didn't have the same concept of land ownership, the English could force them out.

In the early 1600s, it was estimated that there were ten to twenty deer per square mile—a number that didn't overtax the forest food supply.

Oh, Deer

The dwindling of the deer population amid rampant poaching led to the establishment of the first hunting season in 1699.

The wealthiest landowners established plantations. These men didn't want "their" deer poached, and they didn't want a monarchy. By the 1750s, they considered themselves independent thinkers and actors, and they wanted to set up a new society where they were free of the rule of the Crown. They apparently didn't see any irony in the fact that they'd enslaved other humans for over 130 years by then. This desire to be free was particularly true of those who'd settled west of the Blue Ridge. When hunting seasons were first instituted, people who lived to the west were allowed to ignore hunting rules and limits, even though they often took the most deer.

The colonists were also motivated by money. A lucrative trade and export business developed in deer parts that were often obtained illegally, from deer that were poached or killed out of season. The success of this illegal market contributed to the drastic decline of the deer population.

The way the colonists treated the people they enslaved and the native people who lived on the land was mirrored in the way they treated the wildlife and natural resources they found in the new land. People and wildlife were viewed as possessions, means to achieve their goals, or obstacles in their way. The settlers overused resources like deer, and drove out or killed the predators they saw as competition—wolves and mountain lions that, along with the conservative, subsistence-only hunting practices of Indigenous people, had kept deer herds at sustainable levels for thousands of years. The colonists upset the natural balance of ecosystems, and nothing was ever the same.

Deer may also have been stressed by rapid deforestation. When colonists arrived, they cleared forests to plant crop fields. By the late 1800s, agriculture was taking up 80 to 90 percent of the land in some states where forests and grasslands had once stood. And where forested land was no good for agriculture, it was timbered anyway. It seems hard to believe now, but by the early 1900s, deer were almost entirely eliminated from Virginia. The new residents of the land had hunted the once-numerous deer to near disappearance in under three hundred years.

As recently as the 1930s, when stiltgrass was first spotted in Virginia, there were almost no deer left in the northern Piedmont region at all. For the next few decades, some states actually *imported* deer from states where they were still abundant. Virginia reintroduced more deer this way than any other state. It still took forty more years for deer to bounce back in the northern Piedmont. By 1980, the region was considered to have reached its capacity for deer—and there were fewer than half as many deer then as there are in Virginia now. It wouldn't surprise me if the increased spread of stiltgrass paralleled the rise of the deer population here, in concert with development.

Today there are around one million deer in Virginia, and most of their natural predators are long since the victims of humans or habitat loss. Now, in some places, the large numbers of deer have facilitated the spread of chronic wasting disease, a fatal disease that leaves deer emaciated, slowly killing them. Even so, it would take at least twenty-five years for the disease to substantially reduce their population. To control the problem we've caused, the only option, it seems, is to hunt them, but there aren't enough hunters to do the job. Each doe gives birth to twins, and the numbers grow. Nature is efficient when we let it work right, but we've never done that on any major scale. Now that we know better, we still aren't doing it.

In 2015, the Virginia Native Plant Society issued a resolution on deer management that included the following definitive statement: "After human land disturbance, over-browsing by white-tailed deer represents one of the most serious threats to our native flora, the vegetative communities they comprise, and the many species that depend on them." The society resolved that "regulated hunting is the most effective and efficient tool for managing free-ranging deer populations in most circumstances." They advocated for developing ethics guidance for hunters and for finding ways to make hunting more humane.

Deer have changed the ecology of the forest by creating more open space, altering the soil composition and compacting the soil, devouring the understory, and transporting and spreading invasive seeds. They

Oh, Deer

could hardly do more damage if they intentionally took spade in hoof and planted stiltgrass seeds themselves.

My advisers recommended that I join a DMAP, the state's euphemistically titled Deer Management Assistance Program, in which landowners whose properties are overrun with deer can register and receive state permission for increased bag limits in the form of additional tags, including a greater number of "either sex" days for antlerless deer—days when mature bucks as well as females and immature males can be killed. Hunting rules and seasons are divided among several regions in the state, and some rules are county-specific. Regulations in this county normally limit an individual to taking six deer in a season, no more than three of which can be bucks, and for each of those bucks, you must first take an antlerless deer. Deciphering the hunting rules reminds me of how confused I was when I first learned to play hearts. I don't want to tell you how long it took me to figure out that "earn a buck"—which refers to the requirement to take an antlerless deer first—does not mean the hunter gets paid.

The DMAP's fine-tuned, site-specific hunting controls acknowledge that the deer population isn't uniform across every space in each regulated region. The program's goal is to keep the size of herds under control. As they note in their literature, "Under optimum conditions, deer populations can nearly double in size annually." Yikes.

There are too many deer, and there are too many invasive plants. Clearly, these two problems are related. On the one hand, I like thinking of the mountain as a safe zone for deer. What's more optimum than this mountaintop hideaway where there's plenty of food and water and shelter, and few people are shooting at them? But once I understood what needed to be done to restore native grasslands on the mountain and what managing invasive plants would mean, it wasn't long before I had to give in and allow hunting.

A few years ago, one of my kids told me about a meme. I thought of it as a game—I'm over fifty; I have no idea what a meme actually is, but this one warned, "Pick two to defend you; the rest are coming to kill you." Say you're preparing for a fight. Your "army" consists of two animals you can choose from among a specific list of options, each of which is available in a predetermined number. The list includes, for instance, ten crocodiles, fifteen wolves, three bears, and fifty golden eagles. Each animal possesses a different natural set of "battle" skills. It's their job to protect you and fight off the enemy—which will consist of all of the creatures you didn't choose for your side. You might think fifty golden eagles is a good choice, but you need to consider whether their ability to carry away your opponents or tear them apart will be useful if you find yourself faced with ten thousand rats and five gorillas. (My son tells me to always choose the rats, and my previous experience with rats tells me this is a given.)

Fighting the spread of invasive plants reminds me of this meme. I'm faced with a widely varied set of opponents, each with its own special skill set—the ability to, say, grow new sprouts from its own roots whenever it's threatened, like the paulownia tree; or emit a highly toxic substance that poisons the soil for other plants, like spotted knapweed; or learn to produce seed closer to the ground each time it's mowed—hat tip, stiltgrass. Some of my weapons are effective, while others fail or even backfire, resulting in accelerated growth and reproduction. Some weapons that might succeed pose too great a danger to the plants I'm working to protect. And, finally, I'm outnumbered. My best approach is to aim for many small victories, while not allowing the frustrating setbacks to get me down. I'll take small victories any day over ten thousand rats.

That August, I labored in a spot near my gate, in ninety-five-degree heat, pulling three-foot-tall stiltgrass to free a patch of goldenrod. A few weeks later, invasive garlic mustard had moved in and taken over. Garlic mustard will be my next project, I guess. The best time to pull it is in early spring.

Then, in cold mid-November, when in a just world these things should have stopped growing already, I found a tiny baby spotted knapweed plant, that little fucker, sprouting from a hairline crevice in a boulder. My game of weed whack-a-mole continues.

Chapter 10

THE BAD TREE

... in which a tree escapes a garden and conquers the world

One day in June, I arrived at the meadow and knew immediately that something was different. The sweet smell of the mountain air, the scent, I'd often imagined, of all the plants around me gently exhaling, was supplanted by another odor. A new plant must have flowered, but what? And a strange phenomenon: flies gathered around my car, massing on the hood. At no time before had I encountered flies on the mountain, except when I'd stumbled on the remains of a wild turkey. Puzzling over the blanket of flies, I walked off intent on tracking the smell to its source.

I have a highly sensitive nose; I got it from my mother. Sometimes it's helpful, even lifesaving—like when my mom smelled gas from across a busy road, which led to the evacuation of an entire high-rise condo in Miami Beach, after which a gas leak was discovered in the pool pump room on the far side of the building. But nothing puts me in a bad mood quicker than being stuck in a place with a bad smell, especially one that no one else seems aware of or bothered by. I once spent a long, sleepless night in a rented room trying to trick myself into forgetting its unmistakable (to me) mildewy odor by keeping my nose pressed into a cup of fragrant mandarin orange tea. This new smell on the mountain made my eyes well up and my nose clog. I soon found the culprit, at the western edge of the meadow: a so-called tree of heaven, or *Ailanthus altissima*, in flagrant bloom, its branches laden with tiny yellowy-white flowers, abuzz with insect activity. With flies, that is. They must have been attracted by the smell; at close range, it called to mind rotting perfumed flesh.

I was experiencing firsthand what led residents of New York City to write passionate letters to the editor of the *New York Daily Times* ... in 1855. "I have been exposed to the almost pestilential odor ... [W]ere it not that ... their time of flowering is nearly past, I should at once leave the City rather than endure." In Brooklyn, "nearly every street ... abounds in this filthy tree," wrote another.

Everyone seems to have their own idea of which bad smell it reminds them of. Some compare it with cat urine. The tree's Chinese name translates to the unsurprising *foul-smelling tree*. I counted myself lucky to be experiencing the ailanthus in an open meadow, where distance and space made the smell less oppressive than it would be trapped amid the city's heat-emanating sidewalks and buildings. Both the male and female trees produce an odor, and even the leaves and the bark smell bad—I've heard these compared with dirty gym socks. If only the tree's other negative features were fleeting and offended only the senses.

The ailanthus had been planted all along those city streets to replace native shade trees that had fallen victim to hordes of inchworms, also known as cankerworms, which defoliated the trees and carpeted sidewalks with their unpleasantly squishy worm bodies. The worms had the habit of dropping onto the heads of unsuspecting pedestrians and getting snarled in men's ornate mustaches, according to letter writers. (This was evidently a heyday for letters to the editor complaining about nuisance plants.) I think we can all relate to our waxed mustache tips catching on

The Bad Tree

insect webs and trapping worms. The city in its wisdom removed those trees and planted the ailanthus amid the sidewalks instead.

If you've read *A Tree Grows in Brooklyn*, the 1943 bestseller by Betty Smith, and one of the most popular novels of the early twentieth century, you may not realize it, but you're already familiar with the ailanthus. It's the titular tree admired by the young protagonist for its ability to thrive in the yard beyond her tenement window. The book's title could be a statement of pride and wonder at the singularity of the tree, as much as it is a statement about the young girl who grows up there. She sees the tree as a symbol of persistence, and it inspires her to persevere in the face of poverty and adversity:

> The one tree in Francie's yard ... had pointed leaves which grew along green switches ... radiated from the bough and made a tree which looked like a lot of opened green umbrellas ... No matter where its seed fell, it made a tree ... It grew in boarded-up lots and out of neglected rubbish heaps and it was the only tree that grew out of cement. It grew lushly, but only in the tenement districts.

This last part was an error. The tree didn't stop at the poverty line; it grew everywhere. Francie may not have seen the trees in Central Park, or in other parts of the city where the ailanthus was planted intentionally or grew by chance. But it was one of few tree species also found in poorer parts of cities, and that, along with the arcing frond-like appearance of its leaves, led to nicknames like Brooklyn palm and ghetto palm.

Like so many plants that became invasive in North America, the ailanthus started out as a desirable ornamental, imported intentionally in 1784 for the estate garden of William Hamilton, a wealthy Philadelphia landowner. Before long, the tree was featured in seed catalogs and marketed to the wider public. A groundswell of interest in plants from Asia made it difficult for some nurseries to keep it in stock, and by the 1840s, the tree's popularity was in full bloom. Its selling points included its speedy growth—customers wanted an "instant" canopy—and its resistance to predators. The tree has no natural predators in North America, but no one would have found that problematic at the time. Soon, the ailanthus was planted for shade along sidewalks and in gardens all over the Northeast. By 1862, over 40 percent of the trees in Brooklyn were ailanthus trees. Those cankerworms weren't much interested in the ailanthus, and as time wore on, neither was anyone else.

***

Until recently, I didn't know there was any such thing as a bad tree. I grew up believing that trees are unequivocally good, and the evidence for this has only accumulated over the years. We all learned it in school: trees give us air to breathe, they provide wildlife with food and shelter, they make shade, and they stop the soil from eroding. If you want to do something good for the planet, plant a tree. All of this I knew by the age of five. Add to that the current understanding of the importance of trees for carbon sequestration. I didn't associate trees with weeds at all. I'd always thought of weeds as fast-spreading herbaceous plants or vines. (And, as I eventually learned, what some people call weeds are actually important native plants; it's all about your perspective. Of course, those people are wrong ...)

A place full of trees is a healthier place. Does it really matter which trees are growing there? Yes, it does. Now I know there are introduced trees that can take over and make the land inhospitable to native wildlife and plants, just like other invasive plants. As much as I'm pro-tree and loath to remove a healthy tree that's merely going about its important photosynthesis business, not threatening my roof or power lines or water pipes, there are some trees whose presence here is anything but innocuous.

A few weeks after we officially became the owners of this land, Adams bush-hogged about a third of the seventy-five-acre meadow. At that point, the vegetation had reached a height of six feet and more in places, with crownbeard leading the way, and black locust and tulip poplar saplings

The Bad Tree

pushing their way into the fields from the wood's edge. I aimed to slow succession—stop the meadow from becoming a forest—while also keeping herbaceous weeds in check. (This was before I consulted experts who told me not to use mowing as a regular approach to managing the meadow.) Adams, for his part, was happy to be let loose on the fields with his bush hog. He thought it might help reinvigorate the old overgrown hayfields. "I might mow a little more than you asked for," he said. "Don't be mad."

Once the vegetation was shorter, I could see the lay of the land. Previously hidden boulders were exposed, poking up out of the hillside. I could see sassafras and locust trees at the edge of the forest, where the terrain began to slope steeply away. I could see what this land was made of.

Not long after, I was walking and daydreaming through the field where I'd been unable to walk before the mow, and imagining what the meadow would look like in a few months, full of wispy golden broomsedge and little bluestem and pink and white wild bergamot and milkweed (not that I could name all those plants back then), when I tripped on a woody plant that stuck several inches out of the ground. This in itself is not unusual, because I'm clumsy, and because the ground here, even after mowing or bush-hogging, is not a uniform grassy pasture, but a network of dips and molehills and bee holes and groundhog holes traversed by vines and punctuated by woodies waiting to trip me up like I'm the comic relief in a buddy film about city folks in the outdoors. Wherever I try to walk here, I also try hard not to fall on my face, for all the normal reasons, and because historically, when I fall I often break something, and I'd prefer to avoid that.

In this case, I only stumbled on a broken sapling about eight inches tall and about a half inch around. The bark was green and speckled, with a spongy, tan pith where it had been cut by the bush hog. Its leaf scars, notches where leaves once were, were shaped like tiny hearts. I recognized that it wasn't a locust or a poplar. That's about where my leafless tree identification skills ended at the time. While I was crouched examining the broken stem, I happened to peer ahead of me, and two feet away there was another broken sapling. And another two feet beyond that, and another beyond that, and another and another ... what turned out to be a long row of these stems that led across the field in the direction of a grove of sassafras at the edge of the woods a hundred feet away. The row was symmetrical, as if these once-saplings had been planted deliberately. But they couldn't have been planted—nothing had been planted in this field intentionally in decades. If seeds had been dropped by birds or buried by squirrels, they wouldn't grow up perfectly aligned. What would germinate in this exact formation? In a creature feature, this would be the moment where the monster gradually reveals the length and extent of its tentacles, and the victim finds she's actually standing not on solid ground but on the creature's back. I imagined the broken stems animated, grabbing my ankles.

Before Celia Vuocolo visited the mountain, there was Justin, an NRCS biologist who would soon move to another position. It was early March, shortly before Adams bush-hogged the field. Justin walked through the tall dried stalks of crownbeard and pointed out the winter remnants of native plants that I hoped I could persuade to multiply. He also noted some of the invasives: paulownia, Japanese honeysuckle, vines of Asiatic bittersweet winding up tree trunks. Back then many of these were new to me. Then, he stopped in the middle of a field where a sixty-foot-tall double-trunk tree loomed over what he explained were its offspring, a grove of skinny saplings sprouting from the roots of the mature tree. The saplings were nearly identical, trunks straight and narrow, each one eight to twelve feet tall and a couple of inches in diameter, their spindly branches reaching identically for the sky. The bark of the older tree—the parent tree—was different from that of the saplings; it was smooth and gray with vertical silvery-white squiggles that reminded me of stretch marks.

Justin stood on a nearby ridge and pointed out more of these trees around the borders of the meadow, but I had trouble telling them apart from all the other trees that were bare for the winter. This one in the meadow seemed to have produced the largest family.

The Bad Tree

"You'll want to get this under control," he said. It was an ailanthus.

That tree stood about fifty feet away from the broken saplings I'd soon find poking up in the bush-hogged field. But I didn't make the connection.

When I first came to this place, I could be forgiven for not focusing on individual trees, when just over their shoulders there was an expansive view of the Blue Ridge. At the time, I didn't think beyond that it was nice to have a few tall trees dotting the meadow, shady oases from what could be an oppressively strong sun. I may have wondered what species they were; I may have even heard their names and not associated them with anything negative. I didn't wonder how they got here. I reasoned that someone had planted them to provide shade for the cattle when this was a pasture. How little I knew.

By April, trees along the edge of the woods around the meadow had begun to leaf out. The double-trunk tree was still bare except for nascent clusters of shiny, dark green sprouts at the tip of each branch. There was nothing familiar to me about its appearance, nothing I could relate to, but nearly everything on the mountain was new to me at the time, so this alone didn't make it exotic. My tree identification skills back then were limited to the leaves I'd pressed into a scrapbook as a kid: maple, oak, willow, pine—don't ask me which kind. If Justin hadn't showed it to me, I might not have known that tree from the numerous black locusts, even with its leaves. The two species share pinnate-style leaflets, but locust leaves are rounded at the end, whereas the ailanthus leaf is longer and tapered.

It turns out that a parent ailanthus tree can not only grow new saplings from its roots, it can send out suckers—new baby trees—as far as fifty feet away. That conga line of broken saplings in the meadow was made up of the big ailanthus tree's progeny. Justin later confirmed that mowing those suckers had been a bad idea. It was best to leave them alone until the right time to treat them with herbicide, because any disturbance would cause them to attempt to proliferate. Now it's hard for me to think of that tree and its "children" as anything but an army of clone soldiers awaiting the order to conquer the meadow.

The tree of heaven is no ordinary plant. It's fast growing, displacing and outcompeting native trees. It can hijack and reprogram the mycorrhizae linked to nearby trees to help it reproduce. Like spotted knapweed, it's allelopathic, and exponentially so: it secretes as many as two hundred different toxins into the soil that kill other plants or prevent them from establishing. Some of these toxins have even been studied for use as herbicides; that's how harmful they are to other plants. These chemicals can also be dangerous to humans—there have been reports of myocarditis in some people who have come into contact with the tree's sap.

The tree will grow almost anywhere, in any type of soil, alkaline or acid, wet or dry, rich or depleted, steep or flat, sunny or shady (though it avoids the deep shade of a closed canopy), or in almost no soil at all, from a rocky, eroded hillside to a crack in a concrete sidewalk, to a crevice on a rooftop. It likes weak, degraded soil the best. It will grow in extreme conditions, like polluted industrial zones with sooty air and contaminated soil, and it will grow in pristine national parks and nature preserves. It prefers open sunny areas, and that explains the success of the expanding grove in the middle of the meadow here. It throws out new shoots from its roots, known as rootlings, clones gathered around the mature tree in an ever-widening circle; and it sends out the distant suckers I unwittingly mowed. In the first year alone, rootlings can grow up to *fifteen* feet tall, and the suckers can make it to ten feet tall.

Considering all the tools in its arsenal, it's not surprising that the tree can easily take over any open area in its vicinity, and its vicinity is now vast. It's difficult to avoid using battle-oriented words when describing the ailanthus, even knowing that in its home territory, in Asia, it's not an undesirable plant (somehow despite being known as a stink tree there as well). Here, things are obviously different. Here, the tree not only threatens from the middle of the meadow and the edges of the forest on the mountain, it lines the highway as I drive toward the mountain. A Virginia Tech road survey over a seven-year period showed that, on average, more than half of every mile of roadway in Virginia held at least one ailanthus tree, and more of them grew along interstates than any other

The Bad Tree

type of road. Sitting in all-too-frequent traffic jams on the highway, I see young ailanthus trees crowding in bushy groves on the median. The study ended eleven years ago; if the survey were done today, it's my guess they'd find few mile-long stretches without ailanthus trees.

Five miles west across the valley, in Shenandoah National Park, the tree has infested hundreds of acres of what should be native oak and hickory forest. The park sits high in the Blue Ridge, and if I stand on this mountain with good binoculars on a clear day, I think I can see the tree of heaven's distinctive canopy and the drippy pink clusters of seeds that can hang on and remain viable even through winter.

One theory about the ailanthus's successful spread into sunlit breaks in dense forests links it with the devastation caused by spongy moth attacks on oak trees. The spongy moth (*Lymantria dispar*, formerly known as the gypsy moth) itself is invasive, brought here intentionally in the 1860s by a French scientist who wanted to cross it with the silk moth. He chose the spongy moth because it's not a picky eater, and he thought it would be less expensive to raise. But he didn't know the two moths couldn't breed with each other; they're not even related. Unfortunately, a few of the spongy moths escaped. The rest is history.

Attacked by the moths, weakened oaks lost their leaves, creating breaks in the forest canopy and allowing light into previously shaded areas. Many of the oaks died off. Nearby ailanthus found their way into these bright spots and took hold. Disturbed soil from construction and clear-cutting created more opportunities for the ailanthus to spread. New roads helped the tree disperse its seeds around states and across the country.

The ailanthus now grows in forty-six states, including Hawaii. It's considered invasive in North America, Europe, the United Kingdom, and Australia. In France, I encountered it thriving on the outskirts of vineyards in Burgundy (*quel horreur!*). I see it everywhere I go—I can no longer enjoy a drive or a hike without remarking on its presence, much to the annoyance of everyone around me. But I want to warn them: the tree must be destroyed, or it could destroy us.

Melodramatic? (Probably. I was a theater kid.) But according to the U.S. Forest Service, out of the 119 most abundant tree species in Virginia, the ailanthus ranks number 37. That means it's more common than fourteen varieties of oak, including the pin oak (which I'd always thought of as ubiquitous), redbud, black walnut, and American sycamore—some of which support the greatest variety of native wildlife among eastern trees. The ailanthus, meanwhile, supports nothing.

The tree of heaven's success can be attributed to sheer excess: besides its excessive pollen that makes people ill, it produces excessive seeds—350,000 in a season, and up to 10 million in a single tree's lifetime. This is not the greatest number of seeds produced by an invasive tree—the paulownia can produce 20 million seeds each year. And producing a large number of seeds can indicate that a tree has low seed viability, meaning not many of them are expected to actually grow into a plant. For instance, the tulip poplar, the prolific, fast-growing native tree, lines the wood's edge, and every year dozens of new seedlings pop up for every one we remove. The poplar produces thousands of seeds each year, but as few as 5 percent of those will become trees. I can only imagine how many seedlings I'd be pulling if the poplars had the seed viability of the ailanthus, which has been documented at *65 to 78 percent*, depending on the age of the tree. And compared with most trees, that seed success decreases only slightly with age. In one study, a 104-year-old ailanthus produced seeds that were 65 percent viable, a rate that researchers called exceptional. As if the ability to germinate at such a high rate and send out distant shoots isn't enough, its seeds must have an almost supernatural ability to stay aloft: seedlings have been found growing as far as two air miles from the nearest tree.

All of these advantages might not matter so much if it weren't for the absence of one important thing: predators. Like other invasives, part of the secret to its success is that the ailanthus hasn't evolved with the local ecosystem, so nothing in the local food web likes to eat it.

Unchecked, the tree will populate this meadow, the edge habitat, the roadside, and any sun-filled break in the forest. If I do nothing, in short

The Bad Tree

order, I'll be left with a monoculture. What once looked to me like a few nice shade trees punctuating these meadows now seemed like looming monsters in a nightmare.

Driving along the primary east-west and north-south corridors through the county, I spot ailanthus after ailanthus along the wooded edges of the roadside. I dream the trees are after me, the spindly branches of the clones wrapping around my neck, as if I were the one rooted to the ground. I awake feeling a sense of futility. When I'm near a grove of these trees, I get an eerie feeling of displacement and discomfort that I can't fully explain. Part of it is their silence. Birds don't sing from their branches. Cicadas don't call. They're silent as death.

Could it get any worse? Yes, it could. The ailanthus developed multiple methods to grow and propagate in order to be successful at home in Asia, where it must foil local insects or accommodate their predation while avoiding permanent harm. Sometimes an insect predator from an invasive plant's home will be introduced intentionally to help control the plant, after careful study, as in the case of the mile-a-minute weevil, which eats only that vine (unfortunately, it doesn't eat enough to kill the plant). Sometimes, not surprisingly, such an introduction occurs by accident. What happens when a "partner" insect from a plant's home country arrives, unannounced and unexpected, in the plant's new territory?

In 2014, a shipment of stone from China hid a stowaway: egg cases laid by the spotted lanternfly were attached to the stone. After hatching at their destination in Berks County, Pennsylvania, the planthopper insect—so-called for its tendency to attack a number of different tree species—proceeded to wreak havoc.

The fly's appearance changes dramatically at different stages of its life span. Early on, it looks like a mutant stinkbug, black with white polka dots. In the next phase, or *instar*, it becomes a giant weevil-shaped ladybug, red with black spots and a tiny pointy head. The final iteration, the adult, is a phantasmagoric moth that fits in your palm, a black polka-dotted red, gray, and white creature, spawn of a Pynchonesque hallucination crossed with a *Silence of the Lambs* movie poster. If you see one, kill it. It spells severe damage for crops, especially fruit trees and grapevines. The fly lays its eggs anywhere it can—tree trunks, rocks, fallen logs, car bumpers, rooftops, lawn furniture. The larvae hatch and eat through the bark of nearby trees to reach the juicy sap. The insect attacks and weakens native oaks, maples, black cherry, black walnut, poplar, willow, pine, and more—so far, seventy species have been identified as spotted lanternfly food. In its early stages the insect eats the young shoots; in later stages it can dine on older plant material. In this way, it depletes nutrients the plants need to thrive. Just as it will lay its eggs anywhere, the fly makes up for its so-so flying ability by hitchhiking: it will grab a ride on anything—your car, your tractor, your bicycle, your jacket, a cord of firewood. That's how it expands its range.

What does this have to do with the ailanthus? Back in its home country, the ailanthus is the insect's primary host. For the spotted lanternfly, this makes any place the ailanthus grows the equivalent of chow time at a welcome wagon. The insect might lay eggs on the ailanthus, it might sip the sap, but it rarely eats enough to kill it—unlike the grapevines it's discovered here. The insect's damage is a threat to the wine industry, our summer fruits, and our backyard plants. This county is grape country, and its wineries are among its most popular tourist attractions, as are many of the 250 wineries throughout Virginia. If people here weren't too concerned about the ailanthus before, they're paying attention now.

Just as there are no native creatures in North America that will eat the ailanthus, researchers have wondered whether any local birds will eat the nonnative lanternfly. An early study out of Penn State seems to show that birds are less likely to eat the insect if it's been feeding on the ailanthus; the theory is that the toxic chemicals in the tree make the lanternfly taste bad, but when the insect *can't* feed on the ailanthus, it could become lunch itself. The chicken was the bird most often observed eating the insect. The praying mantis, many of which are also nonnative, was willing to eat it as well.

The spotted lanternfly's "benefits" include a bonus nuisance: it secretes a sticky, sweet substance called *honeydew*, which may be unremarkable

The Bad Tree

in itself, except that it festers until it spoils and develops a coating of black goo, officially known as *sooty mold*. The honeydew attracts insects that prefer sweets: wasps, yellow jackets, and flies. Because lanternflies tend to mass in one place, the honeydew and sooty mold can build up to an impressive degree. Which means, wherever the insect is hanging out, you'll find honeydew transforming into sooty mold and accumulating on whatever is below: your picnic table, the gas grill, your car, the groundcover under a tree, a tree's leaves and trunk. The flies especially enjoy gathering on rooftops; then the stuff drips down from your overhang, soon turning your patio sticky and black. The more serious problem with honeydew is when it drips onto a plant and the mold develops, sunlight can't make it through to the leaves, the plant can't photosynthesize, and it could die. Even when the plants aren't killed, fruit from trees and vines that are covered with honeydew and sooty mold can't be sold or eaten. Together the invasive host tree and its resident invasive insect are a double whammy with even greater potential to damage native ecosystems and food crops.

One winter, I attended a training session led by the Virginia DOF to learn how to spot spotted lanternfly egg cases. If I find one, I'm supposed to crush it with something (my boot, the handle of my loppers, a sledgehammer). And, I'm supposed to report my findings via an app. Of course there's an app for that! I'm pleased to report that, after an extensive search that included scanning the upper trunks of trees through binoculars, I didn't find any egg cases on the mountaintop, nor have I seen the insect here—yet. (I'm superstitious enough to think I'm tempting fate by even writing that down.) When—if—they do come, the effect could be devastating. Which is one reason I'm so hell-bent on getting rid of the ailanthus trees here while there's still time.

I've never felt so adamant about killing something as I do about this tree and its partner pest. Not since I was trying to rid my suburban yard of a sudden influx of rats from a nearby construction site. I feel a protectiveness toward this land that surprises me. Like most mothers, I spent a lot of time worrying about the various bad things that could befall my kids when they were young. Thinking of each possible threat seemed magically protective; the thing I thought of would not be the thing that happened, so if I could only think of everything ... This land isn't my child, but at times its need and my responsibility for it take my breath away. If I don't stop the ailanthus here—and the autumn olive, and mile-a-minute, and stiltgrass, and spotted knapweed—who will?

We get it, you say. *Cut down the tree and move on already*—right? Wrong. In the olden days, before I knew stuff, I thought all you needed to do with this tree was cut it down and take it away. Now I know that this is, extremely annoyingly, not true. This is, in fact, the worst thing you can do with an ailanthus—even if you treat the stump right away. Why? Cutting it only makes it stronger. Cutting spurs a mature tree to multiply in those many ways I described. The new little trees it sends out will grow even faster than the original, out of spite. What about pulling up the saplings with my reliable weed wrench? The ailanthus stem is soft; clamp those jaws around it, and it breaks right off, which leads to resprouting and multiplying. No wonder I dream that I can't outpace the tree; like the bewitched broomsticks in *Fantasia*, when I crack one in half, ten more, and then twenty, will sprout up.

Even the chemical means of killing the tree are potentially self-defeating if misapplied. The chemical recommended to me for treating ailanthus is called triclopyr. To apply triclopyr successfully with ailanthus, it's best to use a method called *hack and squirt*, and I'm sorry I didn't come up with that term, which sounds vaguely dirty. Hack and squirt involves making a series of small wedge-shaped cuts in the tree's bark, spaced out around the trunk, to create cuplike openings in which you can pour a small amount of the chemical. Two important instructions: you mustn't make a continuous cut around the tree—known as girdling—because then the tree will know it's in trouble and will respond by reproducing itself as described. And, you must apply the chemical immediately after you make the cuts, before the tree gets a chance to begin healing itself. This healing starts as soon as fifteen minutes after the injury occurs, and it prevents, or significantly decreases, the tree's

The Bad Tree

absorption of the herbicide. For this method to be most effective, it needs to happen at the right time of year, too. Try it in early spring, and it will be useless, because the tree is putting out energy, leafing out and flowering. The chemical might even flow back out of the tree as the sap rises. Instead, it should be applied at a time when the tree is drawing energy to its roots. Here in Virginia, that happens in the dead of summer, mid-July through August.

You'll notice when people talk about invasives, they use the word "management" far more often than they say "elimination." I was told to think of ailanthus treatments as a multiyear "plan of attack." But all I can see when I'm in the meadow are these trees of heaven—trees from hell—marching from the wood's edge toward the grasses, and those already dotting the center of the fields aspiring to the wood's edge. If I let them go, before long they'll meet, and nothing else will grow between them.

Remember group assignments in school, how there was always one kid who ended up doing most of the work? Yes, that kid was me. I thought the way to make sure a thing got done thoroughly and correctly was to do it myself. But, as with so many other aspects of this work, I've learned to cede control and rely on others. It's not hard to give in when the project is this overwhelming. But it's also hard not to feel like there is something wrong with me. Shouldn't I be able to manage the equipment and the chemicals and take care of this problem that threatens the meadow? What would Thoreau say about self-reliance? Then I remember that Thoreau took his laundry home to his mother and burned down his friend's cabin. (I know, I know, this is a significant oversimplification of Thoreau's situation, and a cheap joke.)

Every week, I watched helplessly as the leaves sprouted, the flowers bloomed and stank, the trees appeared to grow taller overnight. One hot July day, Brian Morse's crew came out and treated that grove of thirty ailanthus in the middle of the meadow. The following spring, the limbs of the clones reached up to the sky, but they were bare, and they'll never sprout leaves again; however, at least half of the big double-trunk parent tree flowered again the next spring. It's not unusual to have to treat larger trees more than once. Again, I wait. July and August come and go, and unforeseen obstacles mean that the few people who are willing to come out to this rather remote mountaintop in this sparsely populated county are unable to help. The control freak in me kicks in, and I decide to take matters into my own hands.

I obtain sturdy chemical-proof gloves, goggles, and a hatchet. B. is worried about me wielding a hatchet. Because, acknowledged, I'm not the most coordinated person, and no, I've never used a hatchet before. But he knows better than to try to talk me out of ... anything. I go to the hardware store to grab some triclopyr. I need a gallon of it. The clerk in the garden department offers me a spray bottle of 8 percent solution. I check my notes. I need 44 percent, I tell him. He looks at me like if I don't already have two heads, I will by the time I'm done. But I have these chemical-proof gloves! (Which are two sizes too large, so I can't actually move my fingers, because, as I learn again and again, useful work clothes are hard to find in sizes that fit women.)

I know there are intrepids out there who do it all themselves, rodeo around in their side-by-sides from grove to grove toting a backpack sprayer filled with herbicide, a hatchet, and full body armor, but there are too many ailanthus trees here and too much territory to cover. And, reality check: anything strong enough to kill a tree like the ailanthus has to be potent and is best left to the experts, not Harriet Homeowner, the earnest amateur (i.e., me). When I stopped pacing long enough to really think through the implications, I decided it was safest to continue to be patient. I'm not patient; this is one of my many flaws. Yet somehow I chose to throw myself into a project that follows the *slowest possible* timeline.

There are some potential bright spots in the quest to stop the ailanthus, in the form of biocontrols. There's a fungus, *Verticillium nonalfalfae*, that is spread in China by a weevil—the snout weevil, which, I was delighted to learn, looks like the weevil equivalent of an aardvark—and successfully kills the tree. The fungus has also killed ailanthus trees in

The Bad Tree

northern Spain and other parts of Europe. There are no plans to introduce the weevil here yet, but researchers are studying the fungus, trying to ensure that it isn't lethal to desirable plants. It has been submitted to the U.S. Environmental Protection Agency (EPA) for consideration, a process that can take years. This fungus has been a subject of study for more than twenty years, and now Virginia Tech and the DOF are studying it here in Virginia. According to one of the researchers, the lab-produced fungus samples have not proved as lethal as they'd hoped, and they plan to test more powerful strains.

When that study was about to get underway, I considered volunteering my ailanthus population—I want to support the research—but in the end, I didn't do it. What stopped me was the unintended consequences factor. As we've seen, a significant number of plants and organisms that were intentionally introduced through official channels because they were thought to be beneficial ended up causing problems worse than those they were meant to solve. Whenever I read about a species being introduced in a new place in order to combat another introduced species, I think of T. C. Boyle's short story "Top of the Food Chain." In that story, which is about the effects of DDT, the introduction of nonnative species leads to a domino effect that ends with people eating cats. So, I'll watch and wait until science, and the EPA, can assure me that the fungus won't harm anything besides the ailanthus.

I do what I can on my land, but what about everywhere else? It took 250 years from the time the tree was introduced in this country for it to be listed as invasive in Virginia, and many other places still don't list it. Just because most people who know about the tree agree it's a problem doesn't mean enough is being done about it. A newsletter arrived in the mail from the USDA touting the success of their conservation stewardship program on a Virginia farm. It's illustrated with a photo of cattle huddled underneath—you guessed it—a large ailanthus tree.

On a trip to the post office this winter, I noticed a young ailanthus in the parking lot sprouting from a crack in the asphalt. It reminded me of young Francie's persistent tree. It had climbed a chain-link fence, its cluster of rootlings already two feet long and growing horizontally through the holes in the fence, like arms stretching in midair, reaching out, as if the tree were seeking, what? Immortality?

PART III: THE MOUNTAIN OF HOPE

Chapter 11

ON FIRE

... in which fire prevents fire, native plants want to be burned, and I wait for the wind to die down

When I was a kid, people kept trying to set my house on fire. My parents said it was teenagers, the same ones who screeched around the corner in their convertible and tossed cherry bombs onto our lawn in the middle of the night. They were playing with matches, I was told. Yes, they were lighting matches on the stoop at our front door, but I doubt they were playing. We'd come out for the newspaper first thing in the morning and find its charred remains along with spent matches, soot, and burn scars on the concrete. I didn't know why this was happening. But how aware was I at age six? My parents said it was a prank, that they picked our house because it was on a corner and easy to get away. But there were lots of other corners they might have chosen, and other pranks, like ding-dong-ditch, that seemed less dangerous. For whatever reason, our neighborhood teens preferred fire.

On Fire

One summer afternoon, two boys set a fire right in front of me. One shook a spray can and aimed it at a sign I'd made that hung from our lamppost. They ran away as the lamppost caught fire. The burning sign fell, and flames crept up our small patch of lawn toward our house, toward the willow tree and the unusual plants my mother was so proud of, while my friends and I hopped around in fearful glee, as if this was the most exciting thing and the scariest thing that had ever happened to us. My parents called the fire department, but by the time the truck arrived, my father had put the slow-moving fire out with a garden hose. The tips of the grass had blackened, but the grass hadn't burned to ash. Later that night, a policeman showed up at our door, holding a teenage boy by the shoulder. They walked across the singed grass and stood on our char-marked stoop, the boy tousled, sullen, bangs over stricken eyes. My parents called me over.

"Is this the one?" the policeman asked. Not exactly the traditional lineup through a two-way mirror like you saw on TV. It wasn't him.

A few years later, when we moved to a new neighborhood, no one tried to light our house on fire, but my father accidentally set fire to a crabapple tree in our front yard. My father is handy, but, historically, his creative problem-solving often resulted in too much drama. He was trying to rid the tree of tent caterpillars that had moved in on several of its branches. He rigged up a long bamboo pole with a wire hanger attached at one end, and twisted the hanger so it would hold a metal can in which he lit a fire that emitted a steady stream of smoke. He held it up to the webbed tents, like a live-action Statue of Liberty. This did not go well. The crabapple tree's branches caught fire. I don't know if any caterpillars were harmed. The reliable Penn State Extension website warns, in suitable all caps: "CATERPILLAR TENTS SHOULD NEVER BE BURNED WITH FIRE." A well-directed hard spray with a hose would have been wiser. Fortunately, the fire didn't kill the tree. But this incident reinforced a feeling I had after those earlier pranks—that fire didn't belong anywhere other than in our fireplace or at the tops of birthday candles, that it was unpredictable and uncontrollable, that when it was set loose, it became a different sort of creature, like the difference between our pet cockapoo and a wolf.

I've seen lightning strike trees plenty of times, and I know it can start house fires—there was the time a bolt of lightning hit the big poplar next door, zapped our roof antenna, and fried our VCR—but I have a hard time imagining lightning setting a field aflame. And yet, that's exactly what happened on a regular basis when the South was full of prairies. Before any humans came here, many of the grasslands in Virginia and elsewhere in the South were regularly maintained by lightning-induced fires that burned off dead vegetation and helped the native prairie plants go on thriving. Fire was also once a regular natural disturbance in the open fields on this mountain. The old white oak standing in the open meadow had been a lightning victim, as a large scar on the tree's trunk attests. A lightning rod installed sometime after that strike was grounded incorrectly, at the base of the tree, which meant that instead of drawing subsequent strikes a distance away, lightning's damaging power would have concentrated in that spot. The arborist who corrected the problem told me the tree may have been hit by strike after strike over the past few decades. Still, it's been a long time since these meadows burned naturally and regularly on any large scale.

Even if the meadows on this mountain had never burned before, burning matters now. The Prescribed Fire Council, a partnership among organizations including the U.S. Forest Service, U.S. Fish and Wildlife, the Nature Conservancy, the National Park Service, and the Virginia Department of Forestry, published an exhaustive primer on the subject. It begins, "Many of the ecosystems across Virginia's diverse landscape were shaped by the frequent presence of fire over thousands of years. ... Virginia's forests and grasslands are well-adapted to fire, with plant species that have survival or regeneration strategies that not only tolerate fire but some may require it." Burning plants in order to help them grow may seem counterintuitive, but similar to the need to cull deer herds to control invasive plants, preserve the forest understory, and even improve the health of the deer population, it's a step toward correcting the problems our interference has caused. Prescribed fire, or intentional burning, to mimic lightning-ignited fires as well as some types of fires set later by Native Americans, has become an almost standard tool in the restoration process, one of the few tools available that is the same as the one nature would use.

But back in the seventeenth century, people kept starting fires without knowing how to stop them. Precursors to today's fire restrictions can be found in rules from that time governing when intentional burns could take place and how many people were required to work on those fires. On the other hand, fires set by native people in some parts of the country primarily burned the forest understory, keeping things clear around their homes or for forest crop planting, or the fields, to keep them open for the animals they hunted. These kinds of periodic lower-heat fires (as opposed to high-heat fires that can damage large trees) may have burned themselves out faster, too.

During the industrial era, fire became a much more frequent and dangerous event. Fires happened wherever trees were cut down in large numbers, thanks to the practice of leaving behind unusable *slash*, or debris in the woods. Even though the woods here on the mountain may never have been burned intentionally, I wouldn't be surprised to learn that accidental fires related to timbering had happened here, too. Worse, wherever train tracks were laid in forests, locomotives threw off sparks that ignited slash and led to raging forest fires. Engine sparks are thought to have caused a famously devastating event, the Big Blowup of 1910 that burned three million acres in Idaho, Montana, and Washington in only two days. At the time, it was also common practice to burn wood in the middle of a forest to create charcoal. Once those fires were started, they were often impossible to control.

An 1880 USDA survey showed that 95 percent of fires were ignited by white people, and only about a third of those were set on purpose. By that time, any Native American cultural burning practices in the East were already drastically curtailed; most native people who hadn't fallen victim to disease, war, or murder, had been forced to leave their ancestral lands in the wake of Andrew Jackson's infamous Indian Removal Act of 1830, an ethnic cleansing that forced native people to relocate west of the Mississippi.

Although local leaders imposed strict regulations to limit fires, there was no consistent plan until the U.S. Forest Service was formed in 1905. Its founding raison d'être was fire suppression. That overarching goal eventually led to the birth of "spokesbear" Smokey in the 1940s. It wasn't until the 1970s, after scientists began to realize that fires could actually be beneficial, that the Forest Service evolved in its approach and allowed some wildfires on public lands to burn themselves out, instead of actively trying to control them. In the summer of 1988, as in previous years, fires at Yellowstone National Park were ignited naturally by lightning. They were allowed to burn freely at first, and, as predicted, some of these fires burned themselves out after consuming little more than an acre of land. But this approach didn't reckon with the changes wrought by years of fire suppression, which had altered the forest, leaving too much fuel in the form of a highly flammable understory. That, combined with an unusually dry season, meant that most of the fires in Yellowstone not only kept burning, they kept growing, despite efforts to contain them. Firefighters worked to keep the conflagration from endangering people and their homes, but the fires weren't totally extinguished until rain and snow arrived at the park in September. In the end, nearly eight hundred thousand acres had burned. Although the Park Service concluded that its approach to the Yellowstone fires had been ecologically correct, the Forest Service was forced to temper its short-lived "let-it-burn" philosophy in part because of suburban and exurban sprawl. Whole communities had sprung up too close to forests to safely allow out-of-control fires on public lands. Even though the Forest Service now encourages smart prescribed burning, the land is still feeling the consequences of that legacy of fire suppression.

Smokey Bear had been around for almost thirty years by the time people started trying to set fire to my house, and the U.S. Forest Service's most

On Fire

successful mascot taught me that "only you"—me! a kid!—could prevent forest fires. Smokey's face, and the slogan, were stuck on my lunchbox next to Woodsy Owl's "Give a hoot, don't pollute." The metal lunchbox featured a then-popular TV family known as the Partridges who, despite being named after a ground-nesting bird that prefers to live in meadows and at forest edges, did not express an opinion about fire as far as I know.

For most of its history, then, the Forest Service was not known for encouraging people to set fires. As someone who grew up thinking that if I played with matches, I could start a forest fire all by myself, it can be hard to get my mind around this shift in philosophy.

Native grassland and shrubland plants, and some trees, like white oaks, are fire adapted, meaning they must have evolved in the regular presence of fire. Perhaps fortunately, the fescue lawn at my childhood home didn't burn readily, because it's a nonnative plant and not fire adapted. Dead, dry fescue will burn; the green stuff, not as easily. But as my dad accidentally learned, the native crabapple is a fire-adapted, fire-resistant plant, and I don't say that only because of his mishap. The Western Fire Chiefs Association recommends it for planting near houses in fire-prone areas. Other native plants are similarly immune. Blueberry bushes have buds that are structurally shielded from fire; not only will they still sprout, but the year after a burn, the bush will produce even more fruit. Native people in the Great Lakes region have used fire for years to improve the blueberry harvest. Some plants learned to produce a surplus of seeds in response to fire, while others are prompted to resprout from rhizomes underground; their sturdy root system makes regrowth faster and easier even when the surface plant has burned away. And these plants benefit from the nutrients that enrich the soil when vegetation burns.

It may not be surprising, then, that some healthy native grasslands are concentrated around military facilities, where weapons practice can ignite regular fires in the fields. There are rare butterflies that seem to do well living in these simulated war zones, where fires encourage the growth of plants the insects need to complete their life cycle. A population of endangered Karner blues lives on an Army National Guard airfield in Concord, New Hampshire. The regal fritillary, a rare grassland butterfly, is found in few places in the eastern United States; the largest number reside at Fort Indiantown Gap in Pennsylvania, where the butterfly's survival depends on the disturbance of native grasslands brought on by military training exercises.

I've seen what happens when I don't intervene—or when I intervene in the wrong way: I get a hillside full of noxious vines. Once humans have interfered with a place, correcting the problems we've caused requires more interference, not a hands-off approach. As soon as settlers timbered the mountaintop, as soon as the first orchards were planted, they interfered big time in what nature was meant to do up here. That kick-started the disturbance snowball, which grew more intrusive and more complicated and difficult to stop as the decades rolled on.

This mountain land includes over one hundred acres of forested hillsides, and there are hundreds more contiguous acres of forest beyond them on adjoining lands, spilling down the sides of the mountain. When I stand at a high point at one end of the saddle, the mass of distant trees on nearby mountains blurs like finely textured moss. For a long time, we've assumed that the natural end stage in the progression of most landscapes is closed-canopy forest. I can understand why we would think that; our relatively recent history of fire suppression will lead open land to eventually turn to forest, and that's what we find when we encounter most natural, nonagricultural landscapes that have been left alone for a while. But it's not a given; grasslands and savannas throughout the South were once self-sustaining systems, an end in themselves, not a temporary condition. Those landscapes supported wildlife that require resources that forests alone can't provide.

According to Virginia Working Landscapes, grassland and shrubland bird groups have experienced greater population declines than any other bird group in North America. The meadowlark, a grassland bird, has declined by 75 percent. Here on the mountain, I hope to offer a

On Fire

"complete" set of ecosystems—forest, shrubland, and grassland—that will attract and support creatures that need some or all of them, along with those that live their whole lives on one patch of little bluestem or one pokeweed bush.

Grasslands sequester carbon at almost as high a rate as trees—a fact that seems surprising, now that we've been conditioned to think of trees as giant carbon sinks. When forests burn, the carbon the trees have been storing up is released into the atmosphere: bad. In contrast, when grasslands burn, the carbon stays where it is, in the soil: good. With more and more forests burning out of control, grasslands will be a key to containing carbon. Approaching carbon sequestration by planting trees everywhere, or allowing fields like those on the mountain to succeed to closed-canopy forest, is not the easy answer to conservation problems that it seems to be. The solution needs to look more like a prism than a paint-by-numbers landscape.

***

Based on my history with fire, you might guess I'd want to avoid setting these fields on fire on purpose. But I now understand it as a key step; without burning, I'm not sure my restoration plan can succeed. And, I think I owe it to the ecosystem to try a natural approach first. I'm not ready for the alternative, spraying pesticides all over the fields. The land has been subject to enough poisons in the past. Before the 1950s, orchards were routinely sprayed with pesticides that included lead arsenic. Years later, that arsenic can still show up in drinking-water wells and soil. After arsenic came DDT, and then that was banned. We never seem to find out that a substance is dangerous, or being used in a dangerous way, until the damage is already done. (Glyphosate is still widely applied, even though illnesses in farm workers with a long history of exposure have provided the basis for recent lawsuits.) Still, in a place this size, I'm unlikely to be able to manage if I don't use any chemicals at all; instead, I'll carefully spot treat invasive plants using the lowest concentrations that will work.

My reliable team of experts helped me choose three small fields to burn, about six acres of open meadow. I didn't want to burn the whole meadow at once. First, it's important to provide habitat and an escape route for wildlife that leave the field during a burn. And, how would I keep track of all the new growth that would come up after a fire on seventy-five acres? Unlike herbicide, burning won't kill the plant to its roots—and that's part of the point. Native plants will grow back. But like any disturbed area, a burned field sometimes provides an exciting new opportunity for invasive plant expansion ambitions. Rather than being stymied, some invaders benefit from fire. It's particularly risky to burn where my nemesis spotted knapweed is found, because it will take advantage of the newly disturbed open space and proliferate, and then it could leverage its considerable skills to prevent other plants from regrowing. As I'd already seen, weeds can quickly take over despite my best efforts, and it's almost a guarantee that when they're small and surrounded by other plants, they'll escape my notice.

Instead, I'd limit the risk by limiting the area of disturbance and consulting my experts for an educated guess as to what would grow in a newly burned field. The only sure thing is that I'd get more of what was already growing there. I hoped for some good surprises—the more delicate plants whose growth had been suppressed by the competition—and not too many bad ones. Two of the fields I chose were good candidates because of their abundant and diverse native plants and few true invasives, although they had their share of hay grasses and fescue. Nothing's perfect.

One of these fields is a steep northwest-facing slope that falls away below the twin sassafras trees that sit at the top of a ridge. The second and much smaller field, only around half an acre, sits just across the mowed farm lane, and is bounded on two sides by forest. Last spring, B. spotted a pair of mating luna moths there, so I came to think of it as the "luna moth patch." These two fields contain native grasses like little bluestem, indiangrass, purple love grass, and broomsedge, as well as sturdy forbs like goldenrod, milkweed, dogbane, and crownbeard. But thatch buildup in both fields was preventing birds and other small critters from getting around, and it needed to be burned away.

If you saw these meadows in winter, you could pick out the native plants easily, without knowing what they're called or even what each one looks like: point to the dead, dry plants that are still standing upright. The crownbeard is the tallest; next, the dogbane plants with slender bean pods dangling, perhaps a bit of seed fluff still attached; then, the milkweed with dumpling-like pods hanging on; followed by the mixed golden hues of little bluestem, indiangrass, and broomsedge dotting the field; and finally the bald, fragrant seed heads of wild bergamot swaying in the wind. All around those upright natives, the limp, flopped-over nonnative grasses coat the fields in low waves, hiding the soil under dense layers of living and dead matter where only the toughest plants and wildlife can break through.

The third field I chose is the only one that isn't on a slope, and in its content it's a wild card: hardy native plants, but fewer of them, abundant nonnative hay grasses, and a growing mass of native blackberry around the edges. This field was at the greatest risk of filling up with weeds after a burn. Blackberry can spread quickly, too, and that worried me; I didn't want a field full of bramble. Standing at one end, unfortunately, is a large paulownia, which was still alive, despite having been treated the previous summer. It's a savanna-like spot, grassy with a cluster of paulownia saplings surrounding the mature one, and no other trees until you hit the edge of the forest a few hundred feet away. I called that field "the veldt," after a story by Ray Bradbury in which characters disappear into a simulated African grassland in a child's playroom and get eaten by lions. There have been tales of isolated mountain lion sightings around here, so who knows?

I saw these three fields as opportunities to experiment. Would native plants multiply after a burn? Would the blackberry thin out temporarily, allowing other plants to grow? Would I be deluged with weeds?

There are benefits and drawbacks to each season when it comes to burning. Prescribed burns in Virginia are generally performed in late winter—late February through March—the end of the dormant season. A winter burn reduces dead plant matter, which is one of my primary goals for this first burn, and it improves the soil in time for spring growth of (I hope) the native plants waiting in the seed bank or ready to resprout from rhizomes. But woody plants won't burn as well in winter, so a winter burn won't do much to slow the gradual encroachment of the forest on the meadow in the form of hundreds of poplar and locust saplings. In the fields I chose, succession wasn't a real threat—yet. There's some proof that winter burns give shrubs an advantage, and that's something else I'll have to watch for. But burning in winter can clear dead stiltgrass, which otherwise doesn't break down quickly and, like the rest of the thatch, can keep other plants from growing in spring. The fire will unfortunately do nothing to stop more stiltgrass from growing, since it appears in summer along with native plants. I'll have to keep an eye out for that, too.

Some argue that burning should happen in the season when nature would make it happen—that is, in lightning season. But spring and early summer burns are not recommended here, because it's the height of nesting season for ground-nesting birds. A late summer or early fall burn, on the other hand, can knock out saplings and woody plants like blackberry, helping to slow succession, and most birds are finished nesting by then. But temperatures and humidity that time of year are far from ideal for burning. If humidity is too high, everything will be too damp to burn. A fall burn puts native grasses at a disadvantage; once native vegetation is burned off, nonnative grasses can easily take over a field, since at that time of year they will continue to grow. A winter burn it is.

The greatest challenge here is wind; the slightest increase in wind speed or a random strong gust can quickly turn a good burn day into a bad one. And on the mountain, there's almost always wind.

For obvious reasons, there are laws that regulate burning. They differ from state to state and sometimes from town to town. From mid-February until the end of April in Virginia, a 4 p.m. burn law is in effect, meaning you can't burn before 4 p.m. without a permit. The law was instituted in the 1940s because of the frequency of out-of-control fires that occurred in early spring. That time of year poses a particular problem because of all the dead vegetation that has been slowly drying over the winter, combined with lower humidity and higher winds. After 4 p.m., winds often die down, and humidity rises, decreasing the risk. If you want to burn before 4 p.m. within three hundred feet of forest or dry grass (which describes almost everywhere on this mountain), you need to apply for a permit and plan ahead. When I first applied, my burn application had to be submitted to the DOF in the fall for the following winter season.

Driving the rural county backroads I see people burning brush at all times of the day in every season. I'll assume they aren't burning near dry grass or forest. Couldn't a limited brush-pile fire end up spreading? Yes: these accidental burns can keep foresters busy, and any landowner whose nonpermitted brush fire leads to a wildfire is responsible for paying whatever it costs to put it out.

My plan was for the burns to be carried out by the DOF; they have the deep knowledge and experience, the equipment and trained staff. That fall, I was placed on the DOF's list for a dormant season burn. The advice I'd been given was to burn on a three-year cycle, so whatever acreage was burned that winter could be burned again in three years. After three years, a field begins to succeed to young forest, so if I'm trying to maintain early successional habitat, I'd need to burn periodically, not one and done. Farms that participate in conservation programs run by NRCS and USDA are required to burn fields on a set schedule, so they're given first dibs. I learned that some of these folks were still waiting for a burn from the previous year because of staffing shortages and pandemic-related delays. I wasn't part of a program yet (I was still waiting to hear), so my name wasn't at the top of the list, but the DOF burn manager, the forester in charge of planning burns in my region, seemed optimistic about my chances. I showed him the fields I wanted to burn, and he told me to mow eight-foot-wide perimeters as close to the ground as possible to create fire lines, or breaks, that encircled the fields. The breaks were intended to stop the fire from spreading beyond the planned burn area.

In December, Adams bush-hogged neat, level firebreaks around the fields. In January, the burn manager returned and cast a critical eye on the slight but still-visible fescue in the mowed lanes. He told me he needed to see dirt on those breaks. Adams came back and mowed again, this time cutting down to the soil. The cost of all of this mowing was adding up, and even though he did a careful job, it was mildly upsetting to see wide bare-earth paths encircling the fields, almost as if the boundaries had been tilled. The firebreaks constituted yet another disturbance, but there was no choice. I reassured myself it would all be worthwhile when the burn took place, that by spring those dirt paths would feature sprouting native plants, and not (I hoped) garlic mustard and stiltgrass.

Every week, I watched the calendar advance: January ... February ... March ... and no burn. The burn manager suggested I call on whatever superstitious actions I usually rely on to "make" something happen. *Sure, I've got this*, I thought. *I'll just keep worrying about it incessantly.*

One day in late March, I got the call. It was going to happen—I only had to wait a few more days. I let myself get excited about it, then a couple of days later, the burn manager checked in to say he was a tiny bit concerned about potential for extra-low humidity and high winds. Too many dry days, and burn bans might even take effect. The morning of the appointed day, the forester said the odds were now only slightly better than 50:50 that the burn would still happen. If it did, it would be later in the afternoon when the winds would, hopefully, have calmed. There was a burn planned ahead of mine that day, and he'd keep me posted. I hurried to the mountain in time to take the most scenic, if depressing, Zoom call ever—with the civil engineer who was still (several months into what would become a year) working on a plan to fix the gnarly gravel road—and I waited. I watched the smoke rising in the distance from what must have been the other scheduled burn. The mountaintop was calm. But later that afternoon, I got a text. The odds

On Fire

were getting worse by the minute. The wind began to pick up, and my spirits began to drop.

Finally, around 4:30 p.m., it was over. The foresters were still working on the other fire, it was windy, and by the time they could leave the first site, it would be too late in the day to begin, anyway. Deflated, B. and I climbed back into the car. What else could we do? But, there was the consolation of the view west, the silhouetted mountains, and the north-facing ridge a half mile away across winter meadows with their gold spikes of dried bunchgrasses. There was never a bad reason to come here or a bad time to be here.

I drove back to Maryland, dejected, in appropriately slow, frustrating rush-hour traffic. The burn manager still hoped to get to the mountain before the end of March. "Plenty of burn days left," he said. But within a few weeks, time ran out on burn season.

A year later, conditions in the meadows had changed. My application for restoration assistance had failed, by now for a second time; there were lots of worthy projects, and as always, not enough money to go around. On her next visit to the mountain, Celia Vuocolo agreed that, sadly, the invasives had made further inroads, including stiltgrass lining most of the meadows' edges, the unchecked expansion of woody bramble, and mile-a-minute and Japanese honeysuckle vines teaming up to invade new sections of meadow. She implied that, ironically, I might have a better shot at funding now, because the meadow was in worse shape.

By winter, the mile-a-minute vines on the slope below the oak had morphed into heaps of rusty brown, dead plant matter. I was desperate for a solution. The steep slope made it impractical to maintain as meadow, and from what I'd heard, the decades spent trying to wrangle cattle there had been no picnic, either. Between the woods on one side and the mowed path on the other, that field was too narrow with too much open edge area for ground birds and small critters to feel safe from predators on the ground and in the air. The erosion-prone hillside should never have been clear-cut, but that ship had sailed hundreds of years ago.

Celia suggested the best approach to the vine-covered slope was to plant trees there. The field adjoins a forest, and is much better suited to tree cover. I eagerly agreed to this plan. But, first, she told me, I needed to eliminate the invasive vines, because competition from invasives is one of the main reasons tree seedlings fail. Those piles of dead vegetation needed to be gone, or nothing would flourish there except more mile-a-minute.

I could see those mini-haystacks of death pockmarking the slope below the oak tree as I drove by on the narrow, winding county road in the valley, several hundred feet below. I used to fight the temptation to look up at the scenic mountain while navigating the blind hairpin turns. But now I face forward resolutely, in an effort to avoid any accidental glimpse of those tumor-like patches on the slope, the awful reminders of my failure. Over and over I was told there was nothing I could do. Well, it turned out there was one thing.

I wrote to my go-to plant guru, Charlotte Lorick, and asked if she'd ever heard of anyone burning dead mile-a-minute. She wrote back and reported that a colleague of hers had seen mile-a-minute decrease by 95 percent after it was burned in winter. "Burn it!" her colleague exclaimed, with a smile emoji for emphasis.

I'd already met with a forester to get on the DOF list again and try once more for a winter burn in the upcoming season. I sheepishly wrote to him to see if it was too late to add that slope to my burn plan. This may be my best chance to stop mile-a-minute from spreading and infesting the upper meadow. He said he'd come out to see it. This time, DOF would cut the firebreaks the day of the burn, if there was one; that way, they wouldn't be cut unless a burn was absolutely going to happen. (If I'd known that they normally cut the fire lines themselves, using a bulldozer, I would've done it that way the first time, too. But as always, I don't know what I don't know.) At our next meeting, the forester stood with his hands in his pockets and stared grimly at that steep slope, looking like I'd asked him to drive a dozer down, and back up, a 30 percent grade. Which, essentially, I had. But he didn't say no. Then I was left to

On Fire

worry all over again that it wouldn't happen, that another winter would pass without a burn. I tried to train my thoughts in a positive direction. (Ha! Ha!)

The sassafras slope remained part of the burn plan. By some miracle, it was still flush with native plants. But by now, there was an excess of dead, dried grasses, at least a foot of thatch piled up on that hill. The usual directive for eliminating the nonnative grass that leads to that pileup—in this case it was mostly fescue—was to spray it, let it die, *and then* burn to clear out the dead matter, but Brian Morse told me there would be no point in trying to spot spray the fescue. The pesticide wouldn't reach it beneath all the dead stuff; it was better to burn first, a plan I admit I preferred. Then, I could decide afterward whether to spot spray the fescue when it grew back in spring. I marked the burn area boundaries so the fire wouldn't touch an adjoining field that was overrun with autumn olives, because those shrubs are perfectly happy to be burned. I didn't want to make them happy.

I nixed the luna moth patch from the list this time, because of what happened where the fire lines were mowed around it the year before. What had been a pristine spot was now surrounded by invasive multiflora rose and stiltgrass and tough native blackberry and poplar saplings that had shot up and taken over in the bare dirt between the native patch and the woods. Burning wouldn't eliminate those plants and it could allow some of them to spread. It was yet another classic example of the risks of disturbance. Like the mile-a-minute outbreak, it might have been avoidable. Now, I'd need to bush-hog the boundary again— but not down to the dirt this time—and keep it mowed if I wanted to save the little field.

A plume of smoke rose several miles away across the valley, over the Blue Ridge and Shenandoah National Park. An intentional fire had been set in the park: a fifty-acre plot was being burned to reduce invasives. I watched the smoke dissipate as the day wore on, wondering when it would be time to burn here. In the end, I waited a year and a half for someone to light this mountain on fire.

BAD NATURALIST

***

A field on fire is not what I'd imagined. Done safely and correctly, it's not a conflagration of leaping flames, but more like a low carpet unfurling slowly and steadily across the meadow. Weather is a wild card. Winds that are too high in a location like mine could send spot fires sailing out to adjoining fields, and the whole meadow could end up engulfed in flames. On the mountain, the usual flatland maximum fifteen-mph windspeed is lowered to ten. The optimal temperature is below sixty degrees Fahrenheit. The burn boss, the forester in charge of the burns that day and the same man who'd looked askance at that steep slope, explained that humidity is even more important than temperature, and different landscapes require different limits. In most cases, they aim for humidity higher than 25 percent, and here they were hoping for at least 35 percent to prevent the fires from spreading too easily over dry fields. If the humidity creeps up higher than 55 percent, which would be unusual here in winter, nothing much will burn.

The burn boss continuously monitors the climate on the mountaintop. The humidity that day was in the safe zone and high for winter, 35 percent. It was also a relatively warm day for March, fifty-three degrees in the morning, and climbing. The strongest wind gust clocked in at 7.9 mph—so far, in the clear. Wind isn't only a driver for fire, it's a distributor of smoke. The burn boss wanted to avoid thick smoke covering the winding road below and causing potentially deadly low visibility. I later learned they'd placed smoke warning signs on the four-lane highway, more than a thousand feet below, in case the smoke blew down from the mountain and settled on the road; luckily, that didn't happen. I'd informed neighbors in advance so they wouldn't worry, and the DOF alerted the local fire department. Some neighbors later told me they had seen the smoke on the mountain.

The burn crew is a team of eight, a mix of ages and genders, including foresters, firefighters, paramedics, and trained, experienced volunteers.

On Fire

They arrive early in the morning, and preparations begin. The burn boss fills them in on the details of the fields to be burned, how the burn will be carried out, and who will be responsible for what. Two people in the crew would use drip torches to light the fire from different spots around the edges of the field. Others wield leaf blowers to help clear debris from firebreaks or blow burned debris back into the burn zone. More people want to help start the fire than want to lug around the heavy leaf blowers (can't say I blame them). All the crew wear fire-resistant safety clothing, hats, and glasses, and carry backpacks that can open into firesafe shells if they're caught in the middle of a dangerous flare-up. There's a brush truck with a big tank of water in its bed standing by in case of emergency. While the crew prepares, the burn boss continues to check the stats on a gauge he loops over a low tree branch.

Before they cut any fire lines, I ride down the oak slope with the burn boss in a side-by-side to review the burn areas. Coming back up the steep hill, the vehicle labors, sputters, coughs. I mentally prepare to jump out before the thing can give up and roll backward down the hill. I don't think we're going to make it up, but we do. (Some vehicles can handle these hills, and some can't; I cross that model off my wish list.)

A bulldozer operator begins scraping the fire lines. His progress is slow and laborious because of the steep grades. I begin to see an advantage to cutting the firebreaks in advance—creating the lines takes hours, and the more breaks that are already cut, the more fields they'd have time to burn. At one point the dozer driver continues cutting beyond the planned burn boundary, riding out into the middle of tall crownbeard, until I wave him down. Afterward, he says it's easy to get disoriented in the meadow, and I agree. The foresters are being extra cautious, and the fire lines are far deeper than the ones Adams cut the previous year: uneven, gouged earth with chunks of sod and clods of dirt piled alongside the wide trails. In the end, I tell myself, what matters is that eighteen months after I first signed up, the burn will finally happen.

The first burn, on the sassafras slope, doesn't begin until noon, but it goes off without incident. They start the fire near the top of the ridge, against the southwest wind. Fire travels uphill quickly, and if they started burning from the bottom of the slope, they could lose control. I watch from a distance, where they told me to wait. The hillside is on fire; smoke swallows it, and I can no longer see the twin trees atop the ridge.

After a few more hours of preparations, they're ready to burn the oak slope. This time, I'm allowed to get closer to the action, and the air is sharp with the scent of struck matches and burning straw. There's a dramatic flare-up about fifteen feet high when the fire hits a mound of dead mile-a-minute piled on a sapling. Not long after, a vine that climbed up a tall snag at the edge of the woods catches fire, and the tree begins to burn. The foresters act quickly and calmly, cutting a gap in the barbed-wire fence, taking the burning tree down, and working to extinguish the fire before it can spread to the forest. They succeed, but it takes them until late in the evening. There's no time to burn the other half of the slope, even though a long fire line has already been cut there.

After the burn, the fields are charred and blackened; the soil is dark and rich, like the healthiest topsoil I've ever seen. Other than singed blackberry canes that stand up like parentheses in the stark landscape, and the occasional fur of scorched fescue, the field is bare. But not for long; soon, new plants will begin to grow.

No one knows for sure what will come up in the spring when the fields recover. I hope I'll see native plants return in force. I hope those plants will prevail over any invasives that may have been hiding from view or waiting in seeds. One conservation biologist told me not to expect miracles. After a burn on a field he was restoring, invasive Japanese honeysuckle took over. He had to spray everything and repeat the process. I hope for better luck. It reminds me of an old game show, *Let's Make a Deal*. The host would offer an audience member the chance to choose one of three closed curtains on the stage, and they'd win whatever was behind it. Only prior experience watching the show provided any idea of what to expect. Sometimes there was a fancy vacation or a new car behind the curtain, but sometimes there was only a live goat. The soil was like those

On Fire

curtains—who really knew what might be revealed? If I end up with a field of invasive plants, a goat could come in handy.

Smokey Bear's stern face stares out at me from an envelope I receive from the DOF after the burn. Maybe it's time to retire a symbol whose PR campaign was, in a way, too successful for its own good. But I like the bear, and I guess most people do, and that's why he stays, a relic of another time, before we knew what we know now, but not as much as we'll know later. *Only you*, indeed.

Chapter 12

THE MOUNTAIN AND THE VOLE HILL

... in which an ant guards a tree, a lichen becomes a bird's nest, and a vole plants a seed

In his essay "What Is Wildness," David Quammen suggests that the combined features crucial to maintaining a viable ecosystem work the way a heartbeat does to ensure an individual's survival. One of those features is connectivity. "Wildness ... requires living creatures of many different forms entangled in a system of surging and ebbing interactions," he writes. Before I saw Quammen use "entangled" in that way, I was using that word in my head to think about interrelationships on the mountain. It's easy for me to get caught up in the struggle with introduced species here and worth pausing to consider one of the reasons I've undertaken this project in the first place—the delicate connections those species can disrupt. This is part of what has always compelled me about island biogeography; if the mountain itself is an island of sorts, and the meadow is an

The Mountain and the Vole Hill

island on the mountain, the species here must be intricately connected in order to support the mountain community's continued health. Pressures that threaten to disturb those connections could lead to a collapse of the system—a heart attack, or extinction.

From the start, I've been intent on getting to know the mountain by walking and looking and listening. I've realized that in what I think of as my "pre-mountain" life, I've been too apt to ignore or overlook important elements of my surroundings. In my suburban yard, I didn't think to clear the gutters until the rain created an artful waterfall effect in front of my window. And, I didn't realize that the vinca spreading under the poplars was invasive; the flowers are a lovely shade of blue.

On the mountain, I gloss over nothing. I *notice* on a granular level. There's so much here to notice, and I'm at a point in my life when I have time to do it. I'm bombarded with impressions. I try to slow it down, linger over a patch of leaf litter until I spot the camouflaged centipede. I examine enough lichen-covered boulders to (finally) realize all lichens are not alike. I discern the difference between the objectionable odor of blooming autumn olive flowers and the (even more) objectionable odor of ailanthus flowers. I find a patch of wild bergamot and watch the bumble bees move clockwise around each flower, and, later, counterclockwise. When I stare long enough at a bumble bee on a milkweed flower, I see the milkweed bugs, too. Even after years, even once I understand the rhythms and idiosyncrasies of this place, it will still feel new. There's too much here for me to discover all of it. Thinking on an ecosystem level, I wonder if I'll ever be aware of all of the tiny, interdependent parts that make it work.

That first spring, I weaved my way down a steep hill at the north end of the meadow, otherwise known as the sledding hill, passing crownbeard, goldenrod, wild bergamot, pokeweed, and a towering chestnut oak. At the bottom, I found a broad-branching walnut tree, which at the time I could only identify by the mess of old shells scattered beneath. Near the walnut was an unfamiliar tree with smooth, almost shiny bark that resembled the bark of the cultivated cherry trees growing in the orchard.

It was a wild cherry tree, known as a black cherry (*Prunus serotina*). Like many trees that produce stone fruits, almost every part of the tree is potentially toxic other than the fleshy fruit, which, in pioneer days, was boiled and used to flavor brandy. The seeds, pits, leaves, and bark contain an enzyme that produces cyanide when eaten. (One man learned the hard way when he was hospitalized for, on a whim, chewing up and swallowing the pits of sweet cultivated cherries. In a BBC interview after he was treated for cyanide poisoning, asked why he did it, he said, "Curiosity; and you know what they say about curiosity.") The caterpillars that eat the leaves of the black cherry aren't bothered, and neither are the birds, but farmers don't like to find the trees growing around their fields. The dried leaves that fall to the ground are even more toxic than the live green ones, and if cattle or other livestock accidentally eat them, they could die.

The black cherry, however, is second only to the white oak in the number of butterflies and moths whose caterpillars it hosts. More than 450 species hatch and dine on the tree. Compare that with the native dogwood, which hosts only 126 species of caterpillar, and the redbud's 24. It's not as if those trees are unimportant—there are still plenty of species that rely on them, some perhaps exclusively, for food or shelter—but the black cherry's ecosystem benefits are hard to match. The eastern tiger swallowtail visits the tree's flowers for nectar, helping to pollinate it. Then the butterfly lays its eggs on the leaves, and its caterpillars will eat the leaves when they hatch. The coral hairstreak lays its eggs on the tree's trunk, and at night its caterpillars feed on the tree's fruit and flowers. These caterpillars are critical food for hatchlings. Birds like sparrows, tanagers, and some woodpeckers are not only interested in eating the caterpillars that populate the tree; they and other wildlife, like black bears, foxes, raccoons, opossum, deer, mice, and turkeys, will eat the tree's berries and help spread its seeds. The black cherry may make beautiful furniture, but its most important role is the one it plays in the mountain community. Good thing, too, because once I could identify the tree by its distinctive bark and relatively narrow, wavy,

The Mountain and the Vole Hill

pointed leaves, I found it growing everywhere around the meadow, along the forest's edge, around outcroppings, by patches of milkweed. There are even a few intrepid black cherry saplings growing out from under Dominion Rock.

In a TED Talk about the importance of plants, Charlotte Lorick displayed a slide of a bird sitting on a branch. When she asked people what they saw in the photo, they invariably described the bird, never the plant it was sitting on. "We don't pay enough attention to plants," Charlotte insists. She's right. (What was the plant in the photo? A black cherry.)

The tree at the bottom of the sledding hill is the first black cherry I've paid close attention to. There's a reason: there's something odd going on with its leaves. They appear to be infested with a creature I've never seen before: a pinkish wormlike insect approximately eight millimeters long clings to each leaf. Its body is perpendicular, projecting out from the leaf's surface, curving slightly, as if the leaf has sprouted tiny pink fingers frozen in the act of wiggling.

I thought, *This tree is going to die, and there's nothing I can do about it*. Whatever attacked it was clearly too far advanced to be stopped. But, what *was* it?

The good news is the tree's health is not in danger as a result of this insect. Unless the tree is already sick, the black cherry is accustomed to hosting these creatures, much like it's accustomed to hosting those many species of caterpillar without dire consequences. And, actually, these pink fingers aren't worms at all, but evidence of mites. *Eriophyes cerasicrumena* is a mite so tiny it almost can't be seen with the naked eye, and it's unusual in that it has only four, rather than six, legs. Those "fingers" jutting out of the leaves are not part of the insect; they're galls. The descriptively named black cherry finger gall is a hollow pouch that forms when the leaf reacts to being eaten by the cherry leaf gall mite. The mite eventually lays its eggs inside the pouch, and then it dies there. Later in spring, the eggs hatch. The young mites stay inside the pouch until fall, when it breaks open. The female mite overwinters in one of the tree's buds and repeats the cycle.

The mite's continued existence depends on the tree, so if the mite killed the tree, that would be counterproductive. Why would it matter, though, if the mite itself disappeared? Wouldn't the tree be better off? Does the mite serve any useful function? I was convinced that it must; the only creatures I know of that appear to serve no useful function are the mosquito and the oil company lobbyist. And I recently learned I was wrong about one of these: the mosquito is a pollinator. It pollinates a species of orchid found throughout Canada and in parts of the United States, as well as rare orchids that grow in the Arctic. Maybe mosquitoes have another job, too: to cull the herd—us.

I suspected there must be more to the story, and I was right; these mites do have a purpose, and it leads to the sort of entanglement I'm thinking of. There's a butterfly called the cherry gall azure, *Celastrina serotina*. Its larva, the cherry gall azure caterpillar (of course) is a rarity among butterfly larvae: it's carnivorous. The caterpillar gnaws a hole in the cherry gall and eats the mite larvae that hatch there; it even eats the adult mites. The azure controls the mite population on the cherry tree, while using the tree for its own life cycle.

There are numerous species of azure, all with slightly different proclivities, and the cherry gall azure wasn't even officially confirmed and named as a separate species until 2006. We don't know as much about it now as I hope we will years from now. But we do know that when the caterpillar hatches, it's green at first, but then it turns white or pink. Why? One reason that comes to mind: camouflage. It might seem advantageous for the caterpillar to remain green to blend in with the leaves and fool predators. But the cherry gall azure caterpillar may have developed a more effective adaptation. When the leaves are full of pink wormy-looking galls, turning pink when you're a tiny caterpillar may be the best possible disguise. Birds and other insects don't seem interested in eating the galls, so maybe they'll overlook pink caterpillars, too.

This caterpillar has another protector: the ant. There's an ant species that's attracted to nectar that flows from glands on the black cherry tree's leaf stems. The caterpillar itself also emits a kind of nectar, a sticky sweet

The Mountain and the Vole Hill

fluid. The ants come for the tree's nectar and stay for the caterpillar's nectar. Then the ants fulfill their important role, from the tree's perspective: they protect their food source from predators by killing and eating some other insects, including other caterpillars, and they shield the cherry gall azure caterpillar from a wasp that preys on it.

Imagine the effects if one of these relationships were to break down. If no cherry gall azure caterpillar, then the mites could overwhelm the tree. If no ants, then too many insect pests, and perhaps not enough *C. serotina* caterpillars to control the mites. The caterpillars that survive to metamorphose into adult azures go on to pollinate the tree. Now, eliminate the tree, and this whole tiny universe of relationships is split apart. Nothing for the cherry gall azure, nothing for the ants, and perhaps less food for the wasp as well, not to mention less habitat for the hundreds of species of butterfly and moth that lay their eggs on the tree. Some of these caterpillars may eat and reach pupal stage only on a black cherry or a limited number of other plants. Some adult butterflies may be able to gather nectar and pollen only from the black cherry, or a limited group of plants. The eastern tiger swallowtail lays its eggs on trees in the magnolia family, and trees like the black cherry in the rose family. If I went around the meadow's edge and removed all of the black cherries here, it might not eliminate the swallowtail, but it could reduce its population in the meadow, which would in turn reduce the number of caterpillars available to feed the hatchlings of nesting birds.

Specialists like the monarch caterpillar, which famously can only eat milkweed, or a large number of our native bees, which can gather nectar from a very few plants, sometimes only one, are the most threatened by habitat loss. As a specialist of sorts myself, I can relate a little to what it's like being the caterpillar that only eats the insect that's living inside a gall on the leaf of a particular tree.

I've spent a not inconsiderable amount of time trying to find the one magical protein bar that contains only foods I can safely digest. That is, a protein bar without gluten, dairy, eggs, soy, peanuts, or beans, and with an insignificant quantity of sugar, honey, or maple sweetener, or—the latest fad in animal protein substitutes—pea protein. I can eat meat, but even among meat bars, some of these ingredients are invariably present (especially soy). Why bother with protein bars if they're so hard for me to find? Because I need to carry two of them with me at all times, in case I find myself far afield (or literally *in* a field) where there are no safe food options available to me. It's also a given that, once I find the fantasy protein boost that doesn't make me ill, it will be discontinued within six months to a year (where have you gone, Tanka Bar? How I miss you, Carrot Cake Bar ...), or the ingredients will be changed, rendering it off-limits to me, usually by the addition of pea protein.

If people react the way I did at first on discovering the cherry galls, assuming it's a bad insect pest that will kill the tree, and then do what many people would do in that situation and spray the tree, it won't kill the mites, because they're sequestered inside the galls. It will kill the caterpillars that eat the galls and the mite larvae. It will kill the caterpillar of the eastern tiger swallowtail, and any other of the hundreds of pollinator species that use the tree. It will kill the ants that help protect the tree from pests. It's easy to see how this could become a vicious cycle. You spray, and still the galls return, you don't know why, and the tree's health declines because you interrupted the relationships that help support it.

When the balance is upset, the reverberations travel outward. As a human, I have options. I can eat something other than protein bars much of the time. But if pea protein replaces the cherry gall, the caterpillar is out of luck.

Species like *C. serotina* and the black cherry and the mite that rely on each other are part of a fragile web of relationships. There's a term in ecology: *mutualism*. It means that two (or more) different species depend on each other for sustenance, or perhaps for protection, like the ant and the caterpillar. There are all sorts of relationships like this on the mountain, and each one seems to me like a small miracle of interconnectedness. The destinies of species that evolved together over thousands of years are hopelessly intertwined; when one

The Mountain and the Vole Hill

fails, the loss is felt throughout the system. When the cycle is broken, what happens then?

***

I'm following a track made by the wheel of a motorbike racing up the steep slope below the old oak. The anonymous biker's tires have dug gouges into the grassy plateau around the tree. I'm frustrated because this kind of destructiveness can damage the tree's roots, and I wouldn't think of driving over them myself in any motorized vehicle, much less doing donuts around the tree. I can't control much here, but what's the point of owning this land if I can't stop other people from causing damage? *Go kill your own tree*, I mutter into the void.

I walk away from there, past the cemetery and past the hillside, past the twin sassafras trees and down the steep hill below them, headed north. At the bottom, where the path levels off, I come to what looks like another raised tire track of dried mud that cuts across the mowed path. But there's no evidence the dirt bike made its way down the hill to this spot—the grassy hill is undisturbed. And if this raised mud trail was created by a vehicle's tires when someone braked hard or turned sharply, why wasn't there a deep rut on either side of it? I pressed the toe of my boot onto the mud to try to flatten it out, and it caved in under my foot. It wasn't a pile of mud at all—it was a tunnel. That's when I noticed it traveled all the way across the eight-foot-wide grassy lane, from the meadow to the woods, but not in anything resembling a straight line; it curled and undulated, as if a snake passing just beneath the surface had raised the earth in a mold of its own body.

Walking the farm lane with a forester, I point out this tunnel and ask if he thinks it was made by a mole. He isn't sure, but about thirty seconds later, he glances over at the meadow and says, "I think I just saw a vole." Sure enough, the tunnel was made by a vole, not a mole. The vole is a mouselike creature that, unlike the mole, dines almost exclusively on plant material. It has rodent teeth, so sometimes you'll find its toothmarks on shredded bark at the base of a tree. It often exasperates gardeners by eating tender roots and gnawing through seedlings. In the suburbs, the vole isn't generally a welcome visitor in a garden for that reason. As a not-gardener, I never had occasion to be concerned about it. I once thought a vole was inhabiting our flower bed, but it turned out to be a shrew. How to differentiate the vole? It would be helpful to create a chart: vole, mole, shrew, field mouse (and while we're at it, how about a sequel to differentiate groundhogs from muskrats from woodchucks?). The vole is small and brown, with furry round ears, small black eyes, and a furry brown button nose (not pink like a mouse's nose). Its ears and eyes are smaller than a mouse's, and its tail is much shorter, whereas a mouse's tail runs the length of its body.

The shrew, on the other hand, has a long, narrow, pointy snout. It would be willing to eat the vole, but it's harmless to plants. The mole is also carnivorous—it eats insects—and it has oversized front paws with claws for digging. The mole is a valuable help in controlling insects, even if it may occasionally make your lawn look like it's been visited overnight by third graders building science-fair mud volcanoes. The vole on the other hand seems like it's the opposite of helpful. I'm still impressed that the forester could identify a vole from a distance, but I suppose if you're a forester you develop a special sense for animals that will harm trees.

The eastern meadow vole's favorite habitats in Virginia are old fields, open scrub, and early successional fields—well, vole, you've come to exactly the right place! Its diet consists mainly of grasses, which it's got plenty of at its disposal in these meadows, no harm there. It will also eat roots and bark, and it's been known to girdle fruit trees. I squint at the grove of cherry trees and our one ancient apple tree and wonder if they're in danger. The vole will chew right through the skinny stem of a new seedling, leaving a gnawed, pointed top. It particularly likes young trees that are still green. With larger saplings, it strips the bark up to several inches from the ground, as high as it can reach. And voles are attracted to trees that are hidden inside tree tubes, the protective cover that's often placed around new seedlings. The tube is meant to protect the tree from

The Mountain and the Vole Hill

deer browsing and buck rub until it reaches a safer size, but it also provides a cozy shelter where the vole can make the new young seedling into a leisurely meal without being interrupted by, say, a raptor that would like to make the vole into a leisurely meal.

Voles don't hibernate, which is why I saw fresh evidence of their existence in the middle of February. In winter, they can turn carnivore, dining on insect larvae, decaying animals, and even their own young. And they have a lot of young. They can start breeding at only three weeks of age, and keep it up at a rate that sounds exhausting, but it makes sense because only about 12 percent of their offspring will survive past the first month. One litter can yield anywhere from two to eleven babies, and they can do that every month. You do the math. (I mean, I really don't want to do the math.) I don't need to, because in an "up" year in Virginia, there can be nearly *one thousand voles inhabiting every acre*. If there were one thousand voles per acre here, I would know. Right? The place would be teeming with voles—and with the animals that eat them. Maybe it is.

The vole population tends to explode every few years, sort of like mast years in which oaks produce loads more acorns than usual. I'm probably noticing the many tunnels now because this is a "mast" year for voles on the mountain. But why?

Lo and behold, it was a mast year for oaks—the old white oak yielded a bumper crop. The mountain needs more oaks, especially the white oak, which is a keystone plant, meaning that the web of relationships here requires its presence to function and sustain itself, and its absence would cause negative ripples throughout the ecosystem. In his book *The Nature of Oaks*, Douglas Tallamy explains that the oak "supports more forms of life and species interactions than any other tree genus in North America." Pollen microfossil analysis shows that oaks became dominant in eastern forests around eight thousand years ago; species that rely on the oak to survive have done so for a very long time.

When I was a kid, all I knew was that oaks made acorns, and acorns were slippery. They covered the hillside that was my grandparents' backyard in Washington, D.C. At the bottom of the steep hill was a chain-link fence and an alley where everyone put out their garbage cans on trash collection day (that's what many of the alleys in Washington are used for, besides illegal parking). If I wasn't careful while running, I'd trip on the tree's roots, slip on the acorns, and slam full force into the chain-link like a Hanna-Barbera cartoon character, except it would hurt. It may have happened only once, but I'm afraid of falling, so I was habitually careful around those acorns. The acorns were, however, good for: throwing at my brother. And the acorn caps were good for holding between your thumbs and blowing in exactly the right way to produce a loud and shrill whistle.

Clearly, my childhood oak and acorn activities were unlikely to be of much value for the local ecosystem. I had no idea back then what was going on in that tree.

The white oak can host over 550 caterpillar species. We know from studies that one clutch of bird nestlings can eat thousands of caterpillars. Fewer white oaks means fewer birds and fewer pollinators, which means less abundant plants and fewer native plants, since native plants rely on native pollinators. Fewer white oaks also means less food for local wildlife in the form of acorns. As many as ninety species eat those acorns. White oak acorns are sweeter, and therefore preferred by most animals over the bitter, higher tannin acorns produced by the red oak. On the southwest slope of this mountain, there are towering white, red, and chestnut oaks that are around one hundred years old. Red oaks at that age are already beginning to decline, whereas white oaks at one hundred are just getting started.

Since the 1600s, white oaks have been valued as timber. Because they're slow growers, when the forest was cut and then regrew, they were outcompeted for light by the fast-growing red oak and by trees that can take more shade, like the beech. The red oak is also a more prolific stump sprouter, which means when the forest is timbered, cutting both red and white oaks, the red oaks will quickly resprout, gaining the advantage.

But the current red oak dominance may have begun with the demise of the passenger pigeon in the 1890s. The pigeons ate red oak acorns because they germinated in spring when the birds were breeding, whereas

The Mountain and the Vole Hill

white oak acorns germinate in fall. The sheer number of birds could have easily polished off most if not all of the red oak acorns in a nesting area. One nesting area, recorded in Michigan, covered more than eight hundred miles and may have contained over 130 million birds. With all those birds eating red oak acorns, it's easy to see how white oaks could become dominant in forests frequented by the pigeons. But when people wiped out the pigeons, they broke one of the crucial connections among forest species. Far fewer red oak acorns were eaten, and this may have aided the red oak's comeback. (And now, of course, there's the large population of deer that prefer white oak acorns and seedlings.)

The white oak grows so slowly it may not look like much is happening above ground, but a lot is going on beneath the soil. At first, it puts most of its energy into establishing a large, deep taproot that can eventually support a tree that lives hundreds of years. This taproot is also what allows the tree to resprout after a fire, unlike, say, the yellow poplar, which under normal circumstances will outpace the oak. The poplar burns at much lower temperatures, and it will have trouble coming back after a fire. Fire suppression favors the poplar. And, the past seventy-five years have been unusually wet; this, plus fire suppression again, has promoted moisture-loving maple trees over oaks, and in many of the forests that used to be dominated by oaks, maples are taking over.

White oaks are well suited to nutrient-poor soil—soil unimproved by artificial fertilizer. In old agricultural fields that were abandoned to succeed to forest, fertilizer and other inputs may have altered the soil so that it's less hospitable to the oak, and the understory may fill with invasive plants that flourish in the nitrogen-rich soil.

Considering all of these pressures, it's not surprising that white oaks are in decline. The white oak shortage has become so dire that it's impacting a beloved American institution: bourbon. Bourbon is aged in new white oak barrels, which lend the drink all of its color and most of its flavor. Microscopic obstructions in the wood block the pores and make it water- and rot-resistant. You wouldn't want to age whiskey in barrels made of maple. They'll leak, and bourbon aged in non-oak barrels can't even legally be considered bourbon. You wouldn't want to use red oak, either—besides leaking, it won't taste the same. The increasing scarcity of white oak has persuaded the bourbon distilling industry to put money behind planting the trees. They're experimenting with white oak nurseries to grow seedlings in order to produce acorns, in order to grow more seedlings, to try to ensure the tree's future. Trouble is, white oaks grow on their own timeline. They're twenty years old before they even start producing acorns, and they don't hit their peak until after age fifty. (I can relate.) An arborist told me that planting a white oak is a positive step toward a future I won't live to see. He named his daughter Alba, after the white oak's Latin name, *Quercus alba*.

But what does that have to do with the voles? What possible purpose could these critters serve here, especially in such great numbers, beyond serving as someone's dinner? In fields like this one, the vole's most self-evident role is as a popular food choice for predator species like owls, hawks, snakes, coyotes, and foxes. Even skunk, raccoon, and opossum will get in on some tasty vole action. Along with the increase in voles, there will undoubtedly be an increase in these predator species. In fact, that spring I'd see my first American kestrel on the mountain, hovering above the meadow and diving for, probably, a vole.

I soon learned that the vole plays an important role, and not only as a menu item, especially in an early successional habitat that's dense with grasses. Its tiny green droppings are high in nutrients, because it eats its vegetables. When those thousand voles per acre poop in seemingly indiscriminate fashion, they're adding valuable nutrients to the soil. At the same time, their tunneling activities help distribute mycorrhizae, the fungi linked to the successful growth and health of trees. The vole will enter fire-disturbed areas and help jump-start regrowth. Through activities that might look destructive to me—digging and pooping everywhere—voles can alter the direction of succession in a field, and impact the speed at which it occurs. By "deciding" which fungi end up where, voles help choose what will grow. As a side benefit, voles that live in areas that are regularly burned carry more antibodies to dangerous rodent-

The Mountain and the Vole Hill

borne diseases like rabies and hantavirus. How many voles, I wonder, are already at work in my burned fields?

As for all those acorns, at first I thought that voles weren't interested in eating them. I was wrong. Voles love acorns. In the fall, they collect them and hide them, in case they don't have enough food in winter. The vole explosion, then, could be directly linked to the acorn explosion.

Is there a vole problem similar to the deer problem, in that the vole eats too many acorns I'd rather see grow into mature white oaks? Wait, though—I said it collected the acorns; I didn't say it *ate* all of those acorns. When the vole hides acorns, it spreads them far and wide, and (similar to the jay, whose habit of hiding acorns is detailed in Tallamy's book) not only does the vole often forget where it hid them, if one vole finds another vole's secret stash, it will steal those acorns and hide them somewhere else ... and lose track of that hiding place in turn. Many of those lost acorns will germinate and become oak seedlings. Voles will move acorns hundreds of feet, even a thousand feet, from the parent tree. That makes the vole an important distributor of white oaks on the mountain.

And there's more. In winter, if you pick up an acorn from the ground and look carefully, you might find a pin-sized hole. That hole is where a larva has emerged. The larva is the offspring of an acorn weevil, a long-snouted insect that lays its eggs inside an unripe acorn while it's still on the tree. When the acorn ripens and falls to the ground, the larva hatches inside and eats the meat of the acorn at its center. These acorns will never become oaks. Once the larva is mature, it bores its way out of the shell by making that tiny perfect hole. Then it burrows into the ground and waits a year or two to become an adult weevil, when it will return to lay eggs in newly forming acorns, completing the cycle.

The vole thinks these larvae waiting inside the acorns are delicious, and the extra protein boost in winter is a bonus. The vole won't touch acorns from which larvae have already hatched and emerged. But it will collect the ones that contain larvae and save those to eat. In this way the vole helps to control the number of acorn-damaging weevils, moving them far from the parent tree, which in turn helps ensure that more acorns will remain viable and become oak seedlings. Where the vole is found, there's hope for the white oak. The vole is tiny but mighty; it may be treated like a pest when it appears in the wrong place at the wrong time, but it belongs in the meadow.

✳ ✳ ✳

The fortunate vole, to have found itself on this mountain; it will also eat another resource I have here in spades: lichens. Until recently, I hadn't fully separated lichens in my mind from the fungi that signal a dead or dying tree, and when I first noticed their ample numbers, I had to pause a moment to reassure myself that of course most of the trees on the mountain couldn't be old or sick or dying, could they? And for that matter the many thousands of rocks here that are covered with lichens, what about them? The rocks aren't dying, so what's going on?

For one thing, loads of lichens are a positive sign for the mountain. Lichens do best in unpolluted landscapes, and, according to the fascinating doorstop *Lichens of North America*, "to find them in abundance is to find a corner of the universe where the environment is still pure and unspoiled."

It's true that the air is clear and, I've noticed, sweeter up here, even though it's not even half the elevation of Old Rag, whose silhouetted rocky dome I can see from the meadow. It's also true that there are a handful of native plants that grow lavishly here. I've mentioned a few. But there are more lichens than any of those. There are more lichens than rocks here, which is no longer surprising, but I used to think the prevalence of rocks on the mountain surpassed everything else. I felt a kinship when I heard a neighbor who lives on a nearby mountain joke that boulders are her "cash crop." Now it seems that lichens are mine.

Lichens cover the boulders throughout the meadow and the woods; they overlap with each other in their eagerness to take over the low tables of rock that interrupt the expanses of meadow. They appear on snags and up and down the trunks and branches of living trees—oak and poplar in

The Mountain and the Vole Hill

particular, but also locust, sassafras, box elder, hickory, black walnut, and on and on and on. If I park my car overnight, it'll be consumed by lichens. If I stand still for too long, the creatures will grow over me. Okay, maybe only in a Stephen King tale. But seeing the facility with which they take advantage of every available surface, it's not hard to imagine.

And, no, their prolific presence doesn't indicate that anything is dying. It's easy to come to this conclusion because you'll often see them spread over snags or dying trees. Those trees, having lost their leaves, allow more light to seep into a forest, and light encourages the rapid growth of lichens. But lichens don't kill trees; to the contrary, they help promote and sustain life.

Lichens are also soil factories. Some species that grow on rocks induce a chemical reaction that makes rock more soluble, which speeds weathering and breakdown into soil. They collect silt and dust spread by wind and rain. They provide nutrients to the soil, in particular nitrogen, that key ingredient for life. They're literally creating high-quality soil, albeit extremely slowly, over centuries.

I keep referring to lichens as "creatures," in part because they're not plants, and they're not animals, and they're not fully fungi or algae or bacteria, even though they're made up of those. A lichen is a fungus that can perform photosynthesis because it has combined with—or taken over, depending on your perspective—an algae or cyanobacteria (blue-green algae, which, technically, is not an algae ... it's complicated). Fungi contain no chlorophyll and can't produce their own food, until, that is, they find some algae to, eh, *subsume*.

A relationship like this is often thought of as symbiotic, but true symbiosis requires the willing cooperation of the two parties, and it's not always clear the algae is a willing accomplice in this partnership. The fungus contains the genetic material that will result in the transformation of the dual creature into a physical lichen, but the presence of the algae flips the switch on the fungus's genes to make it happen. The fungus controls the appearance of the lichen, but once the two beings combine, they look and behave in a completely different way than they would if they'd never met.

The fungus calls the shots and reaps most of the benefit, while the algae is basically trapped, which sounds more parasitic than symbiotic to me.

I mentioned lichen "species." If you thought a lichen is a lichen, or if, like me, you hadn't given it much thought before, it may be a surprise to learn that there are thirty-six hundred identified species of lichens, and many that are as yet unidentified. When Bert Harris of the Clifton Institute visited the mountain, his attention was drawn by Dominion Rock. He told me there may be as many as thirty different lichen species on that boulder, at least one of which may not have been identified in this county before.

The various lichen species prefer to grow in different places under different conditions of weather, moisture, and chemical or mineral composition. It would be easy to glance around here and decide that all lichens look alike, or that one species must be dominating the mountaintop. But species that grow on rocks don't grow on trees or on soil, and those that grow on granite won't grow on limestone, and those that grow on oak won't grow on poplar or maple, and so on. Even though most of the lichens on rocks here look, from a distance, like the doilies my grandmother used to protect the arms of her sofa, they're not all the same.

Lichens that grow on limestone are interested in collecting lime (calcium carbonate), but those that grow on granite are attracted to rock that contains silicates. The lichens that grow on the many thousands of poplars on the mountain are attracted to the bark of low-acid trees, whereas those that grow on oak and hickory prefer high-acid trees. And one theory holds that the bark of the white oak retains more moisture and attracts lichens that appreciate that feature, too. Smooth-barked trees like beech and black cherry appeal to yet another group. The possibilities for specialization seem endless.

Lichens that grow on bark filter impurities and minerals from the rainwater that runs down a tree's trunk, maintaining the nutrient content and condition of the soil below the tree. They absorb excess water, which helps reduce runoff and prevent erosion. On highly erodible land like this, lichens provide a critical service.

The Mountain and the Vole Hill

There are lichens that grow directly on soil, stabilizing abandoned farmland or disturbed areas, where they're even able to make the soil less reflective so it absorbs less heat and requires less water. Lichens are pioneer species that can live through long droughts. Sometimes they're intentionally distributed on newly burned land to improve the success of tree-planting projects. I knew that some grasslands succeed to shrubland and then to forest, but I didn't know that lichens have their own successional order; colonization of a given area occurs in turn by a series of different species of lichen, until finally mosses and then grasses begin to grow. Considering their vast and varied presence on the mountain, there are few species or habitats here that are not influenced somehow by lichens.

The ruby-throated hummingbird, which has been spotted zipping around Dominion Rock, is the only hummingbird that nests in the eastern United States. It uses lichens to line its nest inside and out, so the nest resembles a lichen on a tree branch. There are insects that use lichens as camouflage, as well. One species of lichen grows directly on the back of the lacewing's larva, which helps distribute the lichen. The lacewing is a crucial predator, controlling the number of plant-eating pests like aphids, mites, and some caterpillars, which also benefits lichens. The lacewing-lichen relationship is 165 million years old, the oldest known association between insects and lichens. Fossil evidence shows that, way back when, the adult lacewing had developed a clever camouflage: its wing pattern closely resembled the appearance of a prehistoric species of lichen.

Slugs and snails eat lichens, as do silverfish and mites. Deer eat lichens that grow on trees. And research shows that lichens make up 80 percent of the warm-season diet of flying squirrels that live in the mountains of West Virginia. I'm not sure how much lichens normally contribute to the vole's diet, but given their availability here, the vole will never go hungry. Some species of lichen that contain blue-green algae are toxic, and these toxic lichen used to be added to bait to poison wolves, so maybe avoid eating lichens yourself, if you're not sure.

Just as the presence of lichens can indicate a clean environment, their relative absence can indicate a high level of air pollution, especially sulfur dioxide, a product of fossil-fuel combustion. Sulfur dioxide in the air can lead to the formation of dangerous particulate matter, which causes heart and lung problems. Improved air quality due to pollution controls in recent decades is leading to the reappearance of lichens in some urban areas. (I wonder whether that will continue, given the increase in large and long-lasting forest fires that send particulate-filled smoke to distant cities.) A species of lichen that was discovered a few years ago in a New York City cemetery had been absent from the city for almost two hundred years. This led the researchers to comment on—depressingly and with a seeming complete lack of irony—"the importance of cemeteries in providing stable green spaces for urban biodiversity." If lichens are an "early warning system" for a decline in air quality, the mountain air here must be extra pure and unspoiled.

* * *

You may have heard people say that siblings tend to fill whatever role is available to them in the family dynamic. Speaking simplistically, where there's a loud, bossy child, you may also find their quieter more introspective counterpart. Where there's a rule follower, you'll find a rule breaker. In my family, I was the "difficult" one. I talked back and acted out. I was assigned no curfew, and there were times when I stayed out nearly until dawn as if to test my parents' flexibility and restraint—if there's no curfew, I'm not late, right? (They weren't amused.) On the other hand, my brother was the delightful, seemingly easy one, good at appeasing our parents because he apparently (irritatingly) agreed with them most of the time. It seemed to come to him as naturally as did obstinacy and contradiction for me. He idolized me, and I defended him when bullies called him names, but in many ways I felt that I never lived up to his admiration. He could be charming; social and talkative, he never met a stranger. His sociability was genuine, even though it was rooted in his

The Mountain and the Vole Hill

obsession about specific subjects, an effect of Williams syndrome. But Williams syndrome also included serious intellectual and physical disabilities, and, as he grew older, severe anxiety and depression, which were exacerbated by another feature of the disorder, black-and-white thinking. For my brother, things were either good or not, with no awareness of nuance. Some of his friends called him "Sunny," because of what seemed a relentlessly cheerful demeanor, but at home, we also saw the flip side.

Managing his life, advocating for his education and all sorts of special needs interventions, was a full-time job for our parents. On paper, he should have been the more difficult one: he needed special schools and special classes, physical and occupational therapy, someone to drive him everywhere, and counseling on daily living skills as he approached a time when living independently as a young adult finally seemed possible.

But almost as soon as my brother reached that long-sought moment of independence, he died by suicide. The natural order of our family was upended. Gatherings assumed unfamiliar shapes; they unspooled in ways I couldn't easily map. There was an empty place in the family, and it threw everything and everyone off-balance. My sense of self tipped. Dealing with the loss of my only sibling, while dealing with my parents' shock and disbelief (to the extent that, for a time, they couldn't accept what had happened and discouraged me from sharing the cause of his death), I didn't have a chance until much later to process and ask myself questions— Who was I if I wasn't his sister anymore? What would our family look like now? A few weeks later, my always healthy, energetic father suffered a heart attack, which his doctors attributed to the stress of my brother's sudden death. Fortunately, he lived. A few months after that, my grandmother, the one who had lived with us since I was a child, died, too. A family unit that had been five was down to three. By then I was married with my own children, but for my parents, their usually bustling, loud, multigenerational household was suddenly reduced to only the two of them for the first time in nearly forty years.

Twenty years later, there's still a hole, a gap in the ecosystem of our family. It's taken me almost that long to slowly shift to try to fill aspects of both siblings' roles, and I do it imperfectly. I can't make up the space my brother left. I can't make up for his loss, but I can try to be more like him when it comes to my relationship with my parents. Pleasing people, for me, is harder than being stubborn. I sometimes wonder how differently my brother and I might have turned out if we could have let go of the roles we created early on, the niches we filled, the way we were juxtaposed against each other and, at the same time, entangled.

When the presence, or absence, of one species can set off a wave of impacts that echoes throughout the web of a specific ecosystem, it's sometimes called a *trophic cascade*. A recent study of Wyoming prairie dogs provides a vivid illustration. A few years ago, prairie dogs in the Thunder Basin National Grassland were attacked by a disease that led to a mass die-off. Predators like raptors, foxes, and badgers, for whom the prairie dog is a staple food, also decreased in number. Then, unusually heavy rains led to unusually abundant grass growth. With no prairie dogs around to graze down those grasses, the result was a drastic decrease in habitat for ground-nesting birds. This in turn led some bird species, like the mountain plover, to disappear almost entirely from the area they'd previously populated. These birds were replaced by a different class of birds that prefer taller grasses. I would think that the impact on the ecosystem from those shifts would also lead to an imbalance in insect life and a changed pattern of seed distribution. That grassland could end up transformed into something unrecognizable.

Humans can be more flexible than most other species. A hole in the family may eventually heal itself, even if it leaves scars. A hole in an ecosystem, like the one left by those prairie dogs, may be filled by an opportunistic species that inextricably changes the system it enters. Will it slip into that niche without incident, a thread woven invisibly into an elegant fabric, or will that fabric be irreparably torn?

Chapter 13

BLACKBERRY FIELDS FOREVER

... in which there can be too much of a good thing

The endless whorls, an arcing barbed prison, but one that keeps danger outside instead of inside. You're a bird nesting beneath a dome of thorns. The sky is canes and silhouetted leaves. It's cool down here, in shadow and variegated light. The soil smells metallic, is all but bare. Scratch and find grubs. Pick and pluck seeds. Your own cave, a kingdom for a tiny bird. Hidden from sharp beaks and talons, safe from canine jaws.

If you're a ground bird in a thicket of blackberry, I, human, am one of the bad elements, barred entry by stiff canes baring their thorns like tiny teeth that bite into my clothes and my skin. Blackberry, hundreds, possibly thousands of square yards of it, is becoming the main ingredient in the shrublands here. When friends in the suburbs hear that blackberry is abundant on the mountaintop, they get excited about visiting and picking the fruit. I tell them they'll need a suit of armor.

When my kids were little, one of them was fond of a picture book called *Jamberry*, by Bruce Degen, in which a boy and a bear rhyme their way through days of whimsical berry-picking. (By "fond" I mean it's one of those delightful books that a parent has been asked to read so many times they begin to plot its disappearance.) Odd that there's nothing in the story about the boy and the bear getting tangled up in bramble that clings to every piece of fabric (or fur) on their bodies, tearing strips out of their clothing and their flesh.

Before I knew anything much about blackberry plants, a day of berry-picking was a highlight of the summer. We'd visit a farm where the plants grew in hedgerows six feet tall and four feet deep. I don't remember noticing thorns while reaching for the juicy fruit, but unless these were thornless cultivars there must have been some. Much of what we picked was polished off in the car on the way home. When I first came to this mountain, I had no idea that the fruit I was excited to see growing here, that I hoped to bake into pies and crumbles, would soon dominate the meadow as well as my thoughts, and that, after two years, I wouldn't have picked enough to bake even a single muffin.

Native blackberry is generally a desirable plant. It feeds native wildlife—birds, bears, deer, mice, turkey (and yes, voles). It shelters ground-nesting birds. Not only do bees use its flowers for pollen and nectar, there are species that will nest in its nearly hollow stems. The bee removes the soft pith, builds cells and stocks them with pollen, and lays an egg in each cell. After hatching, some of the resulting larvae will remain in the stems all winter and mature the following spring. What's not to like about blackberry (besides the ubiquitous thorns)?

Winter should be the best time to walk in these fields. Snakes are brumating (a dormant state that's different from the hibernating done by warm-blooded creatures). Much of the vegetation is dormant, too, or dead, dry and crackling underfoot. There's no wall of green as tall as I am, and I can see where I'm going after months of disorienting meadow blindness. Native plants still stand, for the most part innocuously, the crownbeard a tough but listing beige stalk. To me, this season delivers the meadow at its most picturesque, golden rolling hills against stark cottony skies, and sharp, fresh air; silence but for the occasional sparrow's song and my footsteps. One of the native plants that stands upright in winter is the blackberry. The canes are gathered in quarter-acre patches that sprout up opportunistically throughout the field, no longer surrounded by leafy greens to soften them, to trick me into getting too close.

Winter *should* be the best time. An icy wind cuts from the northwest. I'm wearing the right sort of coat for the weather but the wrong sort for blackberry. The thorns hook onto the fabric and hold, like the plant is intent on stopping me from leaving a party. If I move, they dig in and rend the thin shell of my down coat. I'm stuck. I chop off a tall cane with the loppers, and it falls toward me, clings, hangs on. Down coats, as B. repeatedly points out to me, are made to be warm and lightweight for high alpine sports like skiing, not thorn walking. No coat is totally thorn-proof, but these canes render "ripstop" a marketing fiction. By the time I extricate myself from the field, feathers are popping out on me like I've survived a pillow fight.

By early spring in the luna moth patch, the small native meadow I'd decided not to burn, things were not looking good. Not only was the previous year's eight-foot-wide fireline overgrown with young yellow poplars and bramble, those plants had begun sending scouts into what had been one of the best quality, if also the smallest, grasslands on the mountain. Poplar saplings and blackberry canes stood like outsize giants amid the bunches of little bluestem, purple love grass, and American asters. Yet another example of the consequences of turning my attention to other problems. If I didn't stop them, the vigorous woody natives would soon crowd out and shade out the native grassland plants, and the little meadow would be on its way to becoming a young forest. Poplars officially grow two to three feet each year, but like everything else on this mountain, they grow faster here. Saplings that weren't here a year ago are already three feet tall and it's not even June. I cut them down with my loppers and a cold heart.

Yellow poplars are not technically poplars at all; the tulip-shaped flowers are the clue and the reason they're also called tulip poplars. They're part of the magnolia family. That makes them a host for the eastern tiger swallowtail, but I'm not worried that I'm cutting into the butterfly's habitat; there must be thousands of poplars on this mountain, especially in the younger sections of the woods. Unlike the black cherry, these trees host relatively few species of caterpillar. I have a lot of experience with yellow poplars, since they ring my house in the suburbs. They can reach up to 120 feet tall, and they're weak limbed, which has led to a never-ending ritual of collecting broken branches from the yard after every windy day. They also tend to rot invisibly and keel over without warning. They don't make good firewood, either, because they burn too quickly. I'll leave the saplings from that small field in a brush pile at the wood's edge, for wildlife.

Next I decide to try to make a dent in the blackberry population in the little patch. Both blackberry species I've found on the mountain are already represented here. One bends and dips enough to snake along above the ground, grabbing my ankles as it throws out canes and taking root wherever it touches down, and the other stands forbiddingly upright, waist high or more, daring me to cross it. Between the two, I amass a collection of thorns on my sleeves and my trousers. My gloves aren't up to the job, and more than once I stop to yank them off and pluck thorns from my hands.

The blackberry has so effectively taken over many of the meadows that I'm pleasantly surprised when I find a hidden spot that's sufficiently free of its barbs, where I can walk through without cutting a makeshift trail as I go. I can still traverse a section where a Virginia dogwood stands on a high hill above the old spring, and I watch a dozen turkeys peck around the water's edge below. But in most places, the bramble is closing in; the last time I tried to check on a cluster of new oak seedlings I'd found, I couldn't get anywhere near them. Blackberry had risen up what seemed like overnight and filled the path.

Blackberry Fields Forever

I see two varieties of blackberry here; the one we generally think of as wild blackberry is Allegheny blackberry (*Rubus allegheniensis*), which is endemic to the United States, in particular the eastern United States, and is fond of higher elevations. The name "Allegheny" comes from the mountains and the river. Allegheny is also my dog's litter name. All of the pups in her litter were named after rivers, and she was Allie, for short. When the pups are old enough to leave for their permanent homes, their new owners usually change their names, and those litter names are forgotten. My dog is easily distinguished by her silver snout and the mischievous glint in her eye that says she stole your hat again; the Allegheny blackberry is easily identified by its stem, which is thick and reddish brown, and if you rub your fingers around it, you'll feel ridges—but watch out for its large, sturdy thorns. Its unobtrusive white flowers are small and informal. I never understood describing flowers as informal, but now I get it. Like the crownbeard's raggedy yellow flower, the blackberry flower puts on no airs, but when yards of the plants are in bloom all at once, it looks like the aftermath of a parade, white confetti sprinkled across the meadow.

Each flower contains more than one hundred pistils, in which the ovaries are located. Blackberries are technically self-pollinating, but if they're not also visited by bees and butterflies, like the orange sulphur, that ensure the majority of their ovaries are pollinated, the fruit will be malformed, stunted. The Allegheny's erect canes will eventually grow long enough to curl over and touch the ground, where they'll root and resprout. *Cane* is the name for the thorny stalk that can grow as thick as my thumb. The fruit ripens to dark purple in late summer. If you can fight the thorns to get to it, it makes good jam and pies. But, although it's growing outrageously on the mountaintop, I rarely get a taste. Anything within reach is eaten by local wildlife before I can pick it. And the yield isn't what it should be, considering the broad area covered by the plants. The first year I was on the mountain, I noticed some of the blackberries had been attacked by rust, a fungal disease that appears as an orange coating on the leaves. When the rust spores are released, they settle on other susceptible blackberry plants and into the soil. That means any blackberry plants that grow where there are rust spores can fall victim to the disease. The first year, I spotted it on only a few plants, and I dug them up. Last year, I didn't find any rust at all, but this year, as soon as the blackberry started leafing out, I saw it in places I'd never seen it before. Within a few weeks, it had disappeared, but I'm guessing it's not gone.

The other dominant type of blackberry on the mountain is actually black raspberry (*Rubus occidentalis*). The black raspberry's white flowers and its leaves are similar to those of the Allegheny, but the leaves' undersides are almost white. Its canes are smooth, and, in their first year of growth, they look as if they're coated with a white powder. They're thinner and more flexible, and flop over more easily, traveling almost vine-like through the meadow. The thorns are numerous and multidirectional, but not as tough and daggerlike as those of the wild blackberry; they're more apt to break off and remain stuck to my clothes. The fruits are also purple when ripe, but like red raspberries, when you pluck one of these berries from a plant, it leaves its core behind on the stem, whereas the core of the wild blackberry is edible and stays intact as part of the fruit. Blackberry is a perennial plant; both varieties produce fruit in their second year of growth, then that stalk will die and new ones will grow up. It's not unusual to see first- and second-year growth on the same plant.

Black raspberry produces less fruit than wild blackberry, but its fruit is, in my opinion, sweeter. Not as tasty, though, as wineberry (*Rubus phoenicolasius*), yet another bramble plant in the meadow. Wineberry stems and thorns are deep burgundy, and the plant grows in giant vertical Slinky-like rings that can reach six feet tall. In winter, with its leaves gone, it looks like red razor wire. I don't want to tangle with it. It's related to blackberry, but it's not native. Wineberry was introduced from Asia in the 1890s for cross-breeding with raspberries, but it escaped, and by the 1970s, it had become invasive. I don't want it to spread, but the fruit is sweet as sugar and animals will eat it and distribute the seeds. It's limited to a few small areas, for now.

Blackberry Fields Forever

Between the wild blackberry and the black raspberry, the meadows are full of bramble and should be heavy with fruit by August. But the rust, which can attack both plants, eventually stops them from producing berries. So far, I haven't spotted any telltale orange blobs on the black raspberry, but it may be a matter of time. There is no cure but digging up the plants. Considering the density of blackberry here, that would be a massive undertaking, and it would leave acres of bare dirt ready to be filled by invasives. Once you dig up diseased plants, you're advised to wait four years before planting more blackberries in the same soil. Not that anyone planted these in the first place, and not that I could prevent them from being restocked naturally by the birds and animals that dropped the seeds here to begin with. Where did the rust come from? Another wild area? Another farm? I'll never know.

When Celia Vuocolo wrote the first meadow management plan, she voiced concern about the blackberry. She thought the expansiveness of the bramble, combined with the impenetrable nonnative grasses and infiltrating invasive vines created too many obstacles for the wildlife that could most benefit from the blackberry's presence. Wherever it grows, blackberry quickly forms thickets up to five feet tall and expands in breadth until it hits paths that are kept mowed. There are thickets larger than a quarter acre, and where a thicket has not yet formed, single canes sprout up diligently, setting the stage for more. These masses act as platforms for invasive vines, like Japanese honeysuckle, that use them to climb and expand across the meadow. Early on, Celia had recommended mowing the blackberry in long, wide strips in winter and treating sections of it with herbicide in early spring when the plants start to grow back. The goal was to leave hedgerow-like stands of blackberry, and create large, relatively open lanes between them, almost like the farm where I used to pick berries as a kid. But that aspect of the plan ended up taking a back seat to the pressing problems caused by plants that are actually invasive. And, I balked at the broad use of herbicides on what was, after all, a native plant frequented by native insects and wildlife that could also be harmed.

Michael Pollan writes that America's penchant for lawns has taught us that "with the help of petrochemicals and technology, we can bend nature to our will." It's both an observation and a complaint I can get behind. I worried not only about herbicides but about surfactants, too; these chemicals that help herbicide stick to a plant's surface can be even more toxic than the herbicide itself. Surfactants are largely unregulated, which means that their full chemical composition isn't even revealed by most labeling. There's enough research on the subject to fill another volume. My plan remains to use the bare minimum of the safest pesticides when it's the sole reasonable option; with many of the invasive plants, little else will work.

I asked Celia about goats—don't goats eat blackberry plants? Wouldn't that be an eco-friendly approach? Yes, goats will eat the blackberry, along with everything else inside the (hopefully very secure, like maximum-lockdown secure) fenced area. They'll eat all of it down to the ground. And then it will grow back, because it's not dead. Now, if they repeat this hard browsing enough times, it will eventually exhaust the plant so it stops growing. But I don't have any goats, and I don't have a well to provide water for them, a prerequisite for "borrowing" a herd of goats, or any other livestock.

By year two, when Celia revised her plan, conditions had worsened on the mountain. Whether because of human activity, or deer and other wildlife distributing seeds, or a combination of factors that favored the success of invasive plants, the invasives were doing what they do best: proliferating. There were too many threats to the meadows to expend significant energy (and money) dealing with a plant that, after all, actually belongs here. Celia advised that the blackberry would have to wait for another year, or two. She rewrote the restoration plan to focus solely on tasks aimed at controlling invasive plants, for now. On the plus side, she thought the blackberry thickets might keep back the stiltgrass, which had gathered its forces and now surrounded most of the meadow, waiting for an opening to slide into its heart. And I couldn't help thinking of the native bees whose larvae may be overwintering in the blackberry canes; that seemed like a good reason to avoid cutting a field of blackberry in winter.

But several months later, in the spring following the burn, it was clear that the blackberry posed a pervasive threat; it had expanded to such an extent that it could no longer be ignored. The meadow was in danger of becoming one giant seventy-five-acre blackberry bramble.

This is going to be the never-ending story here. Between temperatures, precipitation, the combination of wildlife, human disturbance, and what's hiding in the soil, I can count on many species, even some of the native ones, growing and expanding beyond the boundaries of their usefulness for the existing ecosystem, until the meadow can't support them without transforming itself. That transformation may not always be what I'm working toward or hoping for.

Brian Morse took one look around the mountaintop and said he knew I was getting a lot of conflicting advice on the blackberry, but it was time to mow it; I shouldn't let it go any longer. "You want a grassland, but you're on your way to a shrubland," he said, pointing at one broad field of what looked like mostly blackberry. "I saw or heard all sorts of early shrubland birds here," he told me. "Prairie warbler, orchard oriole, common yellowthroat—great shrubland birds." And he was sure there were already bluebirds nesting. But he'd heard none of the classic grassland species. I had to agree. I'd hoped to hear meadowlarks on the meadow, or spot a short-eared owl or the elusive grasshopper sparrow; I was still waiting.

There's nothing wrong with a shrubland; it's an important native ecosystem and a normal stage of succession. It's beneficial transitional space, that feathered edge between forest and grassland. But if I let things go the way they're headed, I'll lose the plants and wildlife that thrive in an early successional grassland, and possibly also those that need both grassland and shrubland for their life cycles, like the bobwhite. And the blackberry threatens to become a monoculture. A healthy ecosystem requires biodiversity; any monoculture threatens that balance.

In late July, once the ground birds finished nesting, and the blackberries finished flowering—but before they had a chance to produce fruit—that was the time to do it, Brian advised. If I could get to it before it fruited, there would be fewer seeds to spread to new areas, and mowing would reduce its dominance and provide pathways for birds and wildlife. If I mowed different patches each year, that might help control the spread. The plants mowed in July would grow back, just as they would if goats had eaten them. But they wouldn't have time to reach their full height and they wouldn't flower again this season. Sure, blackberry could still spread by rooting from those flopped-over canes, but at least I could remove one avenue of reproduction. It would be a start.

Back when I first started this project, I was instructed to mow a portion of the meadow each year. Then, months later, I was told, oh no, absolutely do not plan to manage this land by mowing it. It will clear the way for invasives and leave dead plant matter on the ground, blocking native plants. Confused yet? I was. Even once I'm able to make a decision, having weighed all of the competing values, I'll continue to agonize over whether it was the right one.

But Brian wasn't concerned about creating thatch. At this point, he said, "it's worse not to mow." He was far more concerned about a mountaintop of solid blackberry thicket. He advised that I bush-hog the blackberry into islands, rather than strips, reducing these fields of thorns and leaving space between the patches where wildlife could forage and navigate. The idea of islands of blackberry did seem more natural to me, since that's how the plant tends to grow here, radiating out in a mass, rather than naturally forming a row.

But if I mow it in the middle of July, even if the ground birds are done nesting, I may kill other creatures that can't move out of the way quickly enough. It's like the trolley problem, but with nature.

"Sometimes to solve a problem, you have to break a few eggs," said Brian.

What about box turtles? I wondered.

They'll burrow, I was told.

Native plant restoration on the mountain is a little like writing a book. There are enough people who will tell you their preferred approach, what

Blackberry Fields Forever

works for them or for writers they've taught. But it might not work for me, because of the way I'm wired, the paths my brain takes to get to a sentence or a word or an idea, or the way I perceive connections and structure.

There is no single approach, and I don't know if anything I try will stick. There may be a "best" way, but it depends on whom you ask, and that way could change between now and next year, depending on what happens in between. Each expert knows a way that will work, or has worked, but they can't say for sure whether it will work on this mountain, or in this particular field, with this particular combination of plants and soil and weather. No one knows for sure what will grow here if the seed bank is given a full chance, beyond WYSIWYG—what you see is what you get. The experts can speak from their experience, though, and their experience is far greater than mine, so I listen like it's the law.

Adams remembers chastising me early on regarding the importance of recapturing the old hayfield. It was regularly hayed before falling to some years of neglect. Now, after nearly three years of working with me, he's no longer surprised that my goals don't include encouraging a hayfield and its nonnative cool-season grasses. He knows that when I'm talking about growing "good plants," they're not the same as the plants he'd consider good as a cattle farmer, but he gets that they're good for what I want them for, and he understands why I'm doing it this way, just as I understand why he does things differently on his land. Which doesn't mean we never attempt to convince each other to try it another way.

***

Our "conference room" was a tent around seven feet tall at its center, tall enough to stand up under it, and with open sides for the occasional breeze. We'd pitched it under a mature locust grove, surrounded by crownbeard, to take advantage of the shade. We set up a few plastic lawn chairs and used it for meetings and as a place to escape the changeable extremes of weather on the mountain. The tent was home to caterpillars with spiky hair that would cocoon themselves into its folds, as well as mice, a variety of alarming spiders, and the occasional walking stick. Then one summer, the conference room went missing. Wary of the wind, which had already snapped two sets of tent poles, we'd collapsed it and left it staked with the chairs resting on top. A few weeks later, the tent had mysteriously disappeared, although the chairs were there leaning against a rock. I texted Adams and asked if his guys had moved the tent when they mowed. But they hadn't seen any tent, only the chairs. How strange. I climbed up on Dominion Rock, and peered around at the fields. I could see nearly the whole meadow from there, almost to the oak tree. But there was nothing—not a scrap of fabric, not a fragment of pole, no evidence the tent had ever existed. It was a mystery. Until about a month later: in early fall on a walk to the south gate, we peered into a steep ravine that had been invisible from the rock, and there was the tent. It was trapped in bramble. B. and I picked our way down the hillside and worked to extricate it, managing to gingerly slide out a couple of the poles, but the tent itself was entrenched in a blackberry thicket that was fronted by multiflora rose, a thorny invader whose stems were so heavy and tough my loppers glanced off them like I was cutting concrete with kid scissors. The tent was glued in place, a bug in a web; whatever we tried to do to release it seemed only to ensure it became further enmeshed. In the end we gave it up. The mountain, and the blackberry, won again.

Chapter 14

LET IT GROW

... in which good things come to those who wait

Ever take on a project so exciting and exhilarating that it seemed to take over your life overnight and reframe everything you thought you knew about yourself as a person—the idea that you could do something big and ambitious, something that would have an impact that was different from anything you'd tried before? But once you got into the nitty-gritty heart of it, you began to worry you weren't equal to the challenge? That all the work you'd done so far would amount to nothing? You poured in sweat, blood, tears, dealt with the aches and pains and various mixed metaphors, only to hit a brick wall?

This is how I was feeling when I thought the burn I had by then waited almost eighteen months for was *never* going to happen. And if it didn't happen, how could I make any progress at all on the mountain? Little did I know then that the burn *would* happen a month later. But that's where I was in my head the morning I was scheduled to meet with Michael Shaw: My project was a failure. I was a failure. I was a late-middle-aged woman who had trouble remembering vocabulary words she'd known since grade school, with a bum hip and nothing to offer but a hillside covered with weeds. I couldn't even drink caffeinated coffee or, even better, a whiskey to lament my recent misspent life. Why did I think this project was a good idea? Who needs this frustration? I could be at home staring at my unfinished novel! Then I'd at least feel like I was accomplishing something.

A few hours after my morning reckoning, on what turned out to be an alarmingly warm seventy-degree day in early February, Michael came out to the mountain, and we wandered the hills and fields to talk about the possibility of planting trees on the slope below the old oak, as Celia Vuocolo had advised. Michael works with Friends of the Rappahannock (FOR), a nonprofit focused on protecting rivers and streams throughout Virginia by way of a daunting array of programs, including education and ecological restoration. When I heard that they help with restoration projects, I contacted them the old-fashioned way, by filling out their website form. I soon heard from Michael, who manages projects in six counties, and he made plans to meet me.

That day, I told him I felt like I'd hit a dead end. I'd been waiting and waiting for a prescribed burn, and even if it happened, I'd run out of ideas for how to pay for everything else that needed to be done here, that seemed to need doing all at once. But he was unfazed. Unlike people I'd met from other organizations, Michael didn't warn me about staffing shortages or competition from numerous other projects that needed funding. He just said, "This is doable; this is going to happen." I wasn't sure I'd heard right. "People want to plant trees," he told me. "There's plenty of money available for that."

They'd plant three hundred seedlings per acre, he explained. FOR would monitor them periodically and replace those that weren't doing well. They guaranteed an 85 percent survival rate. To put that in context, I'd visited farms after tree-planting projects in which a similar number of seedlings had been planted, but only a third to a half of them survived the first couple of years, and the landowners and farm managers were pleased with that outcome. No shade (and no pun), either; it's hard to

Let it Grow

grow trees from seedlings in places like this, where my good friends the deer and the voles are numerous and hungry. But Michael explained that his projects were required to maintain a survival rate of at least 80 percent—and they did. Wouldn't all those trees compete with each other for food and light as they grew? But the DOF creates the planting plan that maps out where and how many trees will be planted, and they expect about half of the seedlings to eventually reach full height and maturity, while the other trees are naturally competed out. They recommend overplanting, because it prevents invasive plants from making inroads into the new young forest. That seemed like an especially good idea on a hillside drowning in mile-a-minute vine.

I'd be able to choose the tree species from a list of natives that would grow well on this type of land, with advice from Michael and a DOF forester. Funding would be secured by Michael's organization. It would be months before anything could be finalized, and, if it worked out, we wouldn't know in time for the fairly short planting window in the spring, so planting couldn't occur until the fall. But the mountaintop was an ideal candidate for tree planting on that slope as well as in a riparian zone around the spring and stream that ran through a ravine in the southwest meadow. "Something will happen here," said Michael. "Probably more than you think."

I left that meeting with a feeling of lightness I hadn't experienced in months of focusing on—brooding about—the mountain. My gloom had lifted, my attitude flipped in only an hour. I kept repeating the words to myself like a mantra: *something will happen here*, until the music swelled and a song started in my head—*something great ... is coming ...* That story doesn't have a happy ending. Will this one? For the first time in months, I think it might.

One month later, the prescribed burns happened. I was elated. Until that moment, I'd felt like a wallflower, watching other farms around me carry out burns, remove invasives, and move forward with their plans, while I waited, eager for my turn but with few apparent prospects. Michael had made it clear that burning the oak slope was an important step in preparing the area for tree planting. The work that had seemed stalled now appeared ready to take a giant leap forward.

Over the next few months, I'd learn whether the burn had been a success, clearing the way for native plants, or a failure, clearing the way, primarily, for noxious invasives. Given the combination of plants present before the burn, my prediction was a mixed bag, but would the results be weighted in favor of native plants, or ... ? I didn't want to get my hopes up. Am I essentially a pessimist, or am I a realist? Or am I a dreamer who usually thinks what I imagine probably won't work out, or am I still writing fictions in my head, or am I just a typical Pisces? (Emerson wrote, "Astronomy to the selfish becomes astrology," so I guess it's better that I don't know much about what a Pisces is supposed to be, but I don't have a solid grasp of the oversoul, either.)

* * *

I'm sitting in a plastic chair under a honey locust tree, eating my lunch. I would be sitting in the shade of the tent, if the tent had not been devoured by the bramble monster that lives in the meadow. I'm eating tuna mixed with chopped celery, radish, and parsley, with oil and vinegar, sprinkled liberally with black pepper. A tiny moth alights on my hand. Its upper wings are tan with darker brown bands at the bottom, but when it spreads its wings, startling bright tangerine underwings are revealed. It's appropriately called the orange wing. I learn that its larvae eat the leaves of the honey locust; no wonder the moth appeared where it did. It flutters up and is gone, or I think it is, but I find it has landed in my tuna salad. Maybe it's curious about an omnivore's diet. I wait but it doesn't flutter back out; it's stuck, its wing plastered against the inside of the container, coated in oil. I try to help, gently maneuvering my fingernail underneath it to give it a boost up and out, but it only ramps up flapping its free wing and sinks into the tuna. All I did was frighten it, and all it accomplishes with the flapping is getting itself further mired. I take a spoon, finally, and lift it out. It's too late.

Let it Grow

Of course I interpret this as a gigantic head-bopping metaphor for my impact here. What if I'm ruining things without even knowing that I'm ruining them? Ronald Reagan infamously said that the most terrifying words were "I'm from the government and I'm here to help." Instead, I offer: *I'm from the human race, and I'm here to help.* As in, everything I do has an impact in tiny ways I may never realize, and in big ways I can't help but notice. I can only try to work in a way that I hope will skew toward a positive impact. How will I know? One way is by looking at a burned hillside.

A few weeks after the burn, the color of the soil on the oak slope resembled my favorite French roast coffee (decaf, but I pretend it's not), and here and there, tiny sprouts started to peek out of the dirt. I was vigilantly watching for any sign of mile-a-minute vine. Farther down the hill, the vines that had enveloped the elderberries the previous summer had been burned off and the grove had not only survived but was thriving, with more than a dozen shrubs starting to green up. I was relieved.

On the sassafras slope, square stalks of crownbeard were rising, along with the ferny leaves of yarrow, clusters of wild bergamot, bunches of deer tongue—a native grass—and the ever-present blackberry growing anew around the previous year's now-fire-blackened canes. Soon I'd see dogbane and pokeweed coming up, too. Unfortunately, I also spotted purple crownvetch, a weed that had been abundant everywhere the previous spring. A few autumn olives were growing in the burned field (it had proved impossible to exclude all of them from the burn area); they'd begun to bloom vivaciously. The sickly sweet odor of the flowers had me choking, my eyes itchy and watering. I made a note to ask Brian if he could treat them whenever it was time. Despite those bushes, most of what I saw in that field appeared to belong there.

After a couple of months, the oak slope looked like a different place. It had exploded with native plants, their stalks striving upward, leaves unfolding: many of the same plants I saw on the sassafras slope—crownbeard, wild bergamot, and milkweed leading the way—along with horse nettle, Carolina geranium, tick trefoil, and poison ivy. Purple dead nettle, a weed that some bumble bees appreciate because it flowers early, was doing well, too, but I wasn't discouraged; the balance of native plants outweighed its presence.

The birds were returning from their migrations, and I began to hear the familiar calls of the previous spring. I understand now why spring is synonymous with hope. After the relative quiet of winter, when the song sparrow and white-throated sparrow seemed to dominate, the fields were coming alive again with a full orchestra of sound. The chiding of the yellow-breasted chat, the *cheereetee* of the common yellowthroat, and the quick interjections of the red-eyed vireo joined the cries of the field sparrow, the goldfinch, and the Carolina wren.

The return I'd been dreading came in May, when the tiny tendrils of mile-a-minute vine began to appear on the oak slope. It sprouted up everywhere, getting cozy with tick trefoil and Carolina rose. If it wasn't killed soon, it would take over the same way it had the summer before. I asked Brian if there was any way to spray it without killing all the native plants that had begun to grow on that hillside. Fortunately, there was; the vine is surprisingly susceptible to low levels of herbicide that, when spot sprayed, would be unlikely to harm nearby perennials. Not long after, he and his crew spot treated the vine on the burned field, and by later the same day, much of it was already dead; the surrounding plants showed no sign of harm. I knew that more of the vine would come up as the season wore on, and I'd have to stay vigilant.

Then something else happened: we heard the first bobwhite call on the mountain. It came from the bird ravine near Dominion Rock. Things weren't perfect here—they never would be—but this was a clear sign that the meadows were moving in the right direction.

I told Brian I hoped to plant trees on the oak slope. He said, "I plant trees, but still I think Mother Nature is a better planter than we are." If trees wanted to grow on the slope—and they did, they wanted to grow nearly everywhere—why not let them grow and fill in on their own?

But Mother Nature is also notoriously indifferent; she was the one who covered the slope with mile-a-minute, and I didn't necessarily trust her

Let it Grow

to grow a diverse forest without help. Michael Pollan wrote, "We've made so many changes in the land that some form of gardening has become unavoidable." Maybe it is a *kind* of gardening—the kind a not-gardener would be willing to undertake—planting a specific selection of tree species, giving them a head start, rather than letting nature decide.

If I don't attempt some redirection, as I already have by burning and weeding and treating invasive trees and vines, what would I accomplish here? A guiding philosophy behind this restoration project is that some level of interference is the only way to bring positive change. I understand that Brian was only trying to warn me that a large-scale tree planting might not be the slam dunk I'm anticipating. (Even planting trees on a small scale can fail, and for unknown reasons, like the holly tree in our suburban backyard that looks frailer every year despite B.'s interventions.) The overrepresented herbivores here will put a lot of pressure on those seedlings. But if I don't plant new trees, the poplars that already have a head start along the wood's edge and on several patches of that hillside will outcompete everything else, as they already have in some of the younger forests here. Like most native plants, poplars can be part of a healthy ecosystem, but they're not a complete ecosystem on their own. If they're allowed to take over the slope, oaks and other slower growing trees won't have a chance.

Not all tree planting is alike or helpful. It's not enough to randomly plant trees everywhere, or to plant a tree farm—a single species of tree, a monoculture for the purpose of commerce, without regard for biodiversity. But in this case, on this slope, planting trees seems like the wise choice, as Celia had explained, and Michael and the DOF forester who wrote the tree-planting plan, agreed.

History is written all over this land. Just ask the mile-a-minute vine. I may at times feel conflicted about making hard choices here, especially since I can rarely be certain of the outcome. When I started this project, the hillside below the oak had been maintained as open land for years, and I thought I'd keep it that way. I didn't know I'd need to plant trees there not only to prevent an invasive vine from taking hold and overwhelming the rest of the meadow, but also because it may be the better choice for managing the steep, highly erodible slope for wildlife. But would I someday regret planting a new forest on open land? This hadn't been my plan, and it would be easy to feel like the plans I'd made were hijacked by a vine. But, really, I want to think the mountain is telling me what needs to happen. I'm at the mercy of history. I can make all the plans I want, but the mountain has its own plan; it will do what it will do, and I'm the one who will always need to adjust.

Even though by then, Celia, the NRCS biologist who'd stuck with my project for two years of planning and advising, had moved on to a conservation nonprofit in another state, her insights continued to guide me. While I waited to hear about the tree planting, I also waited for my cost-share application to go through its mysterious motions. If my latest application was approved, it could help accelerate my restoration work on all seventy-five acres of open meadow over the next three years. (In addition to eventually planting trees on the slope, the NRCS plan would address problems in the *whole* meadow, and this was an enticing prospect.) But their plan would put off tree planting until the end of those three years to first allow for two years of herbicide treatments to reduce competing vegetation on the slope. To accomplish that, they'd require broadcast spraying on the entire steep mountainside, more than a dozen acres of mixed plants, many of which are native. And the spraying wouldn't end there. By the end of the project, 75 percent of the meadow would have been broadcast treated at some point during the three-year cycle. If I didn't get approved, it would be a moot question, but if I did, the cost share would be contingent on strictly adhering to the required procedures and timing for every step of the project.

FOR prefers to manage competing vegetation by bush-hogging and burning, with few or no chemicals, because they're focused on protecting waterways. Michael told me he's successfully planted seventy-five thousand trees this way in the past several years, and he thought bush-hogging the slope before planting would be sufficient without

Let it Grow

large-scale spraying. And, if I worked with them, the trees could be planted a lot sooner.

The NRCS calls on years of science and research to back up their approach to knocking down unwanted plants in old fields. Their methods are proven—but are they right for me, for this place, right now? As author and orchardist Elizabeth Hilborn told me, "You'll have your own truth for your own land."

If I choose Michael's approach, in which the tree-planting area is not treated, I might have to spot spray the invasive vines as they come up again—their seeds are in the seed bank, and they *will* grow back. But native perennials would be allowed to grow around the tree seedlings, and I wouldn't be adding as many potentially harmful chemicals to the soil.

The NRCS biologist who replaced Celia on my project explained that large-scale treatment of the fields—not only the oak slope, but all of the fields—was a necessary part of the process. She mentioned that the meadows wouldn't look great the following year. I'd heard that before; the transition away from old hayfield to a field that's predominantly native isn't necessarily pretty, but if I wanted manicured rolling pasture, or if I still held out fantasies about postcard-perfect fields of wildflowers, I wouldn't be doing this project. What I care about is supporting the wildlife that need a place to live and the plants that feed and shelter them.

"We'll lose the nesting season," she told me, "but the birds will come back."

I didn't find that as reassuring as I think she intended it. The birds were decidedly not coming back in altogether too many places, and weren't pesticides a big part of the problem?

I recall what Brian said earlier about breaking some eggs, but this sounded like enough omelets to feed the Sunday brunch crowd for months. I worry about frogs and toads and salamanders, too, because amphibians are especially sensitive to these chemicals. While I'm thinking it over, a colleague sends me a study by biologist Rick Relyea that examined the impact of glyphosate on frogs. The results are not good. If I spray, the fescue and invasive plants will still undoubtedly return; the frogs might not.

On top of all of this, as I've mentioned, I'm not a patient person. I'm eager to plant trees now, not two or three years from now. Am I being shortsighted? If they aren't sprayed into submission first, will weeds and peppy native perennials like crownbeard, milkweed, and dogbane swamp the small seedlings? Can I spot spray carefully around the good plants, the way Brian did? In the end, I might not have a choice; one option or the other—or both—might not pan out. But I had to be prepared to make a decision.

I began to think that too many people had told me I would have to broadcast spray the fields for that to be the wrong answer. One of the few who hadn't weighed in with an opinion yet was Brian himself, but I expected him to agree; after all, his company sprayed fields like mine all the time. Was breaking those eggs worthwhile?

But I was wrong. At least when it comes to tree planting, Brian advocates a low-chemical approach. He prefers only to spot treat, spraying in a circle immediately around each seedling.

"That way," he said, "we're not destroying any native tree seedling germination—so no mowing after the trees are in either—then we get the benefit of a functioning grassland, including pollinator habitat, even while the trees are establishing."

That helped me decide—as did the news that if I was ready to go ahead, Michael was ready, too. More than a dozen acres of seedlings would be planted in the fall. I'd target invasive plants that crop up among them, but there would be no large-scale spraying on those fields in advance. Would I regret this? Would the seedlings be smothered by the competition? If the folks planting the trees weren't worrying, should I?

In her book, *Water in Plain Sight*, journalist Judith Schwartz writes that "efforts to restore biodiversity are as much about imagination as anything else ... We need imagination in order to have hope and we need hope in order to act." I began this project with idealistic dreams; along the way, I acquired a more realistic—but still hopeful—outlook. I don't see this as a capitulation, but a reasonable tempering of expectations.

One blanket approach *can't* work for all of the different ecosystems on this mountaintop, nor can a solution that might work on a farm in

Let it Grow

the valley, on relatively flat land, be expected to work on a land of rolling and steep hills at a much higher elevation. By drawing from the different philosophies of the experts I've been lucky enough to meet, maybe I'll discover a new way of seeing this mountain, of managing its infinite entanglements. (Maybe that will include a few spotted-knapweed-eating sheep.)

***

The rabbit isn't worried about me yet. It's hunched at the edge of the meadow, nibbling clover. I'm walking along a mowed path, headed toward the cherry orchard on a sweltering July day. It's a rare event to see a rabbit here at all, unlike the suburbs where there's not as much for them to fear and they're as plentiful as squirrels. On the mountain, there are too many predators on the ground and in the sky, and they're usually careful to stay hidden. A smudge overhead becomes a red-tailed hawk, circling, as if to demonstrate. I cross an invisible boundary, and the rabbit freezes. One step more, it hops into the tall grass and disappears.

A week later, I'm walking my dog on the same path. I wish I could let her off leash, but in a place like this, where everything is fascinating, all she'll do if I call her back is turn and give me that droll look of hers that means "you must be kidding." She's smart, but I don't know how she'll react to a snake. And there are a lot of snakes here. She stops to sniff at the spot where I saw the rabbit. She energetically presses her nose into the clover, dashes back and forth tracking the scent, then shoves her face deep into orchardgrass and timothy. I stop her from diving into the meadow. Last time, I didn't stop her, and I spent the evening plucking sticky seeds out of her coat.

Ten days later, I'm walking along the same path, and I feel a body move against the toe of my boot. I instinctively hop over whatever it is, then turn and look back to see a black snake about three feet long and thick as my wrist. I'm surprised to find a snake in the cropped grass, exposed to the midday sun. It raises its head and directs its gaze at me for a long second, like it's wondering the same thing about me. What are either of us doing out here in the heat? It slides into the tall grass in the same spot where the rabbit disappeared, the same spot where my dog had sniffed. It must be a gustatory carnival in there.

Later that day, I return to the cherry grove, but this time I watch my feet for snakes. As I enter the shade of the grove, the path forms an accidental allée, aging cherry trees lined up on either side, their nobby branches hanging low. I hear a high-pitched buzzing, like the whine of a weed whacker; not bees, but flies. They land on my fingers and my elbows, and they bite. The bites burn like bee stings. Tiny gnats dive-bomb my face. That night, I'll find two gnats stuck to my cheek, drowned in the soup of sweat and sunscreen.

I don't often see flies here—there are no cattle on the mountain anymore—but in a moment, I find the source: giant feathers, white-and-brown barred, piled and scattered in a tight circle at the far end of the grove of trees. Some of the feathers were torn out in clumps, and bits of flesh still cling. There was a fight, and the bird lost. Coyote? Fox? If I knew more about those predators, if I could read the signs, I might figure it out. There's so much I still don't know. The feathers belonged to a wild turkey—maybe the loner I'd seen shouldering through a field. The bird's body is gone; nothing is wasted.

The world of the mountain comprises many smaller worlds. In a literal sense, it's an oak-hickory forest and an open meadow of mixed grasses and forbs and a shrubland in the early stages of succession, all of it mediated by the interference of introduced species. The soil is rocky and acidic throughout. There are steep northeast-facing and south-facing and east- and west-facing slopes, some open meadow and some forested. There are softer rolling hills full of grasses punctuated by a few trees, fields that resemble savanna until the bramble closes in. There's a ravine where a spring emerges and a stream runs to the forest. There are hills where granite is buried under a few inches of soil, and where, a few yards away you can dig down six feet (but not much more) before hitting rock. There are boulders full of lichen communities and hiding places for reptiles.

Let it Grow

There's a grove of three dozen old orchard cherries, still standing; a former hayfield, where timothy and rye and orchardgrass mingle with native wildflowers. There's a dry ravine where you'll find little bluestem and American plum trees and doomed ash saplings and some of the most active bird chatter in the meadow—as well as invasive Japanese honeysuckle vines and ailanthus trees. There are rolling fields shot through with thickets of thorny blackberry. There's a small plot with a surprising concentration of native species—purple panic grass, love grass, broomsedge, little bluestem, indiangrass—threatened by intruding bramble and young trees. There's the steep slope below the oak where, amid milkweed and poplar saplings, I found (and lost, and found again) a grove of young elderberry bushes.

By the old cattle gate, there's a tiny pool full of frogs. I hesitate to call it a pond because it's hardly wider than the kiddie pool we used to inflate in the backyard, and not as deep. Unlike a vernal pool, which usually dries up for most of summer and fall, this pool is present year-round. The frogs, toads, and salamanders born here will return to the same spot to lay their eggs the following spring, and on and on, year after year. A steady trickle of water flows into the pool from somewhere inside a wall of earth and moss a few feet above it. Twenty feet across the pool, the water flows away from it, into a narrow channel where it disappears into a hole in the ground that's always lined with fallen leaves. The water burbles and circles that hole like it's a sink drain, but it never stops flowing. Last year, the pool was surrounded by masses of orange jewelweed, tiny delicate flowers that remind me of miniature orchids. Every April, I find egg sacs in the water that look as if someone is blowing bubbles from beneath the surface. Soon after, tadpoles hatch in such large numbers in a small space that at first the water appears black, and then I realize it's crowded with life. I watch them develop over the next few months. One day, a small water snake takes an interest, but the tadpoles are invisible, hiding, I guess, amid the roots of vegetation dangling into the water. I don't know if the snake finds a meal. The tadpoles and the frogs survive, unlike those unfortunate ones I tried to grow indoors all those years ago. This living pond in the "terrarium" of the mountain makes a better home.

This year, a storm deposited a mound of soil at the center of the pond, cutting it in half. I worry the area of still water is too small now, and I won't see many eggs. Come April, I find something I haven't noticed before: the pond appears to be full of plump earthworms intertwined, churning up the sand on the bottom. Could this be a predator? No, I learn there are toad eggs inside the coiled, wormlike jelly. They must have been here last year as well, only I missed them. The eggs hatch, but the tadpoles aren't as numerous as in previous years, perhaps because the pool has shrunk. The jewelweed that was rampant last year hasn't returned; I'm not sure why. But now watercress is growing on the sediment mound and more soil piles up around it when it rains. This watercress isn't native. The late philanthropist Rachel Mellon enjoyed eating watercress sandwiches so much that she asked her staff to grow the plant in one of the ponds on her estate in the Virginia Piedmont. I'd rather have the native jewelweed back, myself. Once the season is over, I decide, I'll dig that sediment pile out, and, I hope, that plant as well, and restore the pond to its previous dimensions. But right away I worry that I might disturb creatures holing up there for winter, that digging out the sediment could cause a different set of problems. I make a note to find out. I'll never run out of questions here.

Adams told me this pond started out as a borrow pit used for fill when the gravel road that leads up the mountain was first cut, decades ago. I never would have guessed. It may not have appeared naturally, but as host to generations of American toad and wood frog, it's more than doing its share, and I'll maintain it. Human disturbance is an inseparable part of what makes this land what it is. I have to choose what to retain and to encourage, what to try to eliminate, and what to let go. Some choices will be easy, like keeping this pond; others won't.

One day in June, I approached the pond and, as usual, I heard the plops of frogs hopping away as they sensed my footsteps. I sat at the edge and watched water beetles skim across the surface, the occasional

Let it Grow

dragonfly touching down for a split second and gliding off again. Something moved in the water where a moment before nothing was visible against the muddy bottom. The creature reminded me of a tiny lobster, with its longish body and disproportionate crablike pincers. Then I realized it was a crayfish. Fifty years ago, I loafed by a stream in some woods looking for crayfish, not believing the kids who said they'd caught them there, and now for the first time, I saw one myself.

***

It's six months since the prescribed burns took place, and the fields are wild with vegetation; you'd never know they were burned. On the sassafras slope, the meadow is tall and lush, with few problem plants other than fescue. A field of goldenrod grows within the field, as does a small field of ragweed; ragweed may get a bad rap because of seasonal allergies, but birds—especially the bobwhite—love the seeds, and it's well contained by the other native plants growing around it. Wild bergamot intermingles with native bunchgrasses like little bluestem, broomsedge, and panic grass. The wind blows across the green field, and everything nods, like it's saying I've actually done something right this time. I monitor the field, watching for weeds by walking the fireline that encircles it; I step carefully because plants are growing there, too, but much more slowly in these breaks that weren't burned. What I see is a predominantly native meadow, to my untrained eye, at least three-quarters native plants. It's only a couple of acres, but it's a start, a microcosm of what I hope to achieve in the rest of the fields.

Will it last? Thickets of blackberry are gathering themselves and beginning to spread out. I cut some of the canes to temporarily keep them in check and give other vegetation a chance to fill in. I pull a few handfuls of bindweed around the edges of the field before it can spread. On the oak slope, it looks like I've managed to get ahead of the mile-a-minute threat. For now. I find a few random clumps that I pull and bag on my own. I know this isn't the last I'll see of it, but it needs sun, and a few years down the road, the new trees that will be planted this fall should cast enough shade to stop it from returning.

At the top of the hill, on the side of the path that wasn't burned, most of the poison hemlock I'd sprayed the previous year hadn't returned, except for one ankle-high plant. Days later, when I'm able to return with a spray bottle, I find that it's shot up like a pubescent teenager—it's over my head, with a stem bigger around than my thumb.

Something here is always threatening to get out of hand, usually many somethings, too many for me to control. I have to stop allowing it all to cloud my vision, or this project will leave me with a constant nagging sense of inadequacy. Will I learn to see the forest, or the meadow, for what it is, or will I always be down here on the ground grumbling at the latest little weed?

***

I walk down the temporary stairs to check out the concrete basement in my house in progress. The wood frame is up, and B. is upstairs wandering the skeletal rooms that are still open to the sky; I hear his footsteps landing above my head. The basement is dark except for sunlight leaking in at the far end, where the room is open to the outdoors. As I descend the steps, I hear a loud sound, as if someone is shaking a box of nails. I can't tell where it's coming from. I call up to B. "What's that noise? What are you doing?"

"Nothing," he says. "I'm walking."

A few more steps, and I'm standing on the basement's concrete floor. I hear it again, but I don't see anything. It's all around me, echoing off the concrete, those nails in a box: TSSH-SH-SH, TSSSH-SH-SH, TSSSSH-SH-SH. If it's not him, I understand there's only one thing it can be: a rattlesnake. I take the stairs back up quicker than I think I've ever moved in my life.

After that, I mean *right* after, we drove to Cabela's, and I bought the first pair of snake-proof boots I tried on.

Let it Grow

***

Long ago when I was sitting under the willow in my yard trying to find a geode by splitting open rocks, I didn't know what I know now. A man who's spent decades studying rocks recently told me, kindly, that I'd had little chance of finding a geode in Maryland; had I grown up in Kentucky, then, maybe. Loads of geodes in Kentucky, apparently.

    I'm glad I didn't know that when I was a kid, because every time, with every suspiciously roundish rock, there was a chance, and what mattered was I thought there was a chance; I was certain of it in my core. I sustained that hopefulness, which may be healthier than the fleeting satisfaction that would have come from actually finding the thing, through all the rocks I'd split open to reveal an inside that looked much the same as the outside. If I never found a geode—and I never did—it was auspicious that sometimes, while I hammered away at the rock, a spark would fly, and the smell of sulfur would sting my nostrils. It was a message from the universe to keep trying. It was enough.

Chapter 15

STEWARD OF SECRETS

... in which I don't know all the answers

When I was well into work on this book, I attended an artist residency. While there I happened to listen to a podcast, which is something I wouldn't normally do at a residency when I'm deep in the writing, but this one struck a chord. *Nighttime on Still Waters* is what its creator Richard Goode calls a "narrowcast"; he broadcasts from a narrowboat he lives on in a canal in England, somewhere in the Warwickshire area. He reads poems, plays music, and talks about life on the canal in soothing, somnolent tones. Listening can feel like taking a long, leisurely bath. In this particular episode, after a walk in the countryside, Goode became philosophical.

"Owning the land and taking ownership of your place within the land are two very different things," he said. Too many people struggle with what he called dysfunctional views of the world, either believing the world and everything in it belongs to them, or "ashamed of the shadow they cast, guilty of the footprints they leave." Goode says these are both the same thing, misapprehensions that lead to alienation, disconnection from nature.

While I'm here on the mountain, however limited my time may turn out to be, I'll be trying to own my place in the land. A friend recently introduced me to the Hawaiian word *kahu*, which can be used to describe a person's relationship with a dog or a cat. I looked it up and found that it explains that relationship as guardianship, or caretaking, rather than ownership.

I don't see how a person can truly "own" land. It's made up of too many elements that are themselves autonomous. Do I own the lichens that are attached to the rocks that are embedded in the earth? Technically, I suppose I do, because I have a paper that humans have agreed means that I do. But no one asked the lichens.

Although at times I still feel like a visitor here, that's my own sense of impermanence talking. Kahu is an apt description for how I see my relationship with the land. I'm a guardian and, in another of kahu's meanings, its honored servant. I can't control the land, but I can tend to it; I can be its watchful keeper, its shepherdess.

Every moment, around every corner, a new discovery waits. What sort of ant is that? Does it belong here? What sort of bee is that? What sort of plant? Why does this grow next to that? Why does it grow here but not over there? I'm immersing myself in the unfamiliar, the way I do whenever I begin a new story—and this is a new story, the story of the mountain, which is both old and new, chapters being written every day, and the story of my life now. There's an old saw that says "Write what you know." I like to write what I don't know yet, what I'm trying to understand. I think that's the key to the mountain, for me, to be open to the questions, the mysteries.

When I started, I knew even less than I thought I knew. The plants I thought were weeds turned out to be native pollinator magnets. The trees I was grateful to see growing in the middle of the meadow, grateful for their shade, were aggressive weeds. Snakes overwinter in old irrigation tubs, native bees live not in hives but alone in the ground, half the plants are toxic, everything has thorns, and on some days the land seems less like a farm than a foreign territory. Which it is, in a sense: an ecosystem, or collection of ecosystems, where I'm an outsider, but where I'm also a part of its world. Though I'm getting to know it little by little, I know enough to know that some of its mysteries will stay that way.

SUGGESTED READING

Books

Brodo, Irwin M., Sylvia Duran Sharnoff, and Stephen Sharnoff. 2001. *Lichens of North America*. New Haven, CT: Yale University Press.

Bushnell, David Ives. 1935. *The Manahoac Tribes in Virginia, 1608*. Washington, D.C.: Smithsonian Institution.

Embry, Paige. 2018. *Our Native Bees: North America's Endangered Pollinators and the Fight to Save Them*. Portland, OR: Timber Press.

Emerson, Ralph Waldo. 1836. *Nature*. Boston, MA: James Munroe.

Emery, Carla. 2019. *The Encyclopedia of Country Living: The Original Manual for Living Off the Land and Doing it Yourself*, 50th Anniversary Edition. Seattle, WA: Sasquatch Books.

Goulson, Dave. 2014. *A Buzz in the Meadow: The Natural History of a French Farm*. New York: Picador.

Hilborn, Elizabeth. 2023. *Restoring Eden: Unearthing the Agribusiness Secret that Poisoned My Farming Community*. Chicago: Chicago Review Press.

Suggested Reading

Jepson, Paul, and Cain Blythe. 2020. *Rewilding: The Radical New Science of Ecological Recovery*. London: Icon Books Ltd.

MacArthur, Robert H., and Edward O. Wilson. 1967. *The Theory of Island Biogeography*. Princeton, NJ: Princeton University Press.

National Park Service, U.S. Fish and Wildlife Service. 2014. *Plant Invaders of Mid-Atlantic Natural Areas: Revised & Updated—with More Species and Expanded Control Guidance*, Washington, D.C.

Noss, Reed. 2013. *Forgotten Grasslands of the South: Natural History and Conservation*. Washington, D.C.: Island Press

Petrides, George A. 1998. *Peterson Field Guides. A Field Guide to Eastern Trees: Eastern United States and Canada, Including the Midwest*. New York: Houghton Mifflin.

Petrides, George A. 1986. *Peterson Field Guides. A Field Guide to Trees and Shrubs: Northeastern and North-Central United States and Southeastern and South Central Canada*, 2nd ed. New York: Houghton Mifflin.

Pollan, Michael. 1991. *Second Nature: A Gardener's Education*. New York: Grove Press.

Quammen, David. 1996. *The Song of the Dodo: Island Biogeography in an Age of Extinction*. New York: Touchstone.

Schwartz, Judith. 2016. *Water in Plain Sight: Hope for a Thirsty World*. New York: St. Martin's Press.

Smith, Betty. 1943. *A Tree Grows in Brooklyn*. New York: Harper & Brothers.

Suggested Reading

Southgate, Emily W. B. (Russell). 2019. *People and the Land Through Time: Linking Ecology and History*, 2nd ed. New Haven, CT: Yale University Press.

Tallamy, Douglas. 2019. *Nature's Best Hope: A New Approach to Conservation that Starts in Your Yard*. Portland, OR: Timber Press.

Tallamy, Douglas. 2021. *The Nature of Oaks: The Rich Ecology of Our Most Essential Native Trees*. Portland, OR: Timber Press.

Tekiela, Stan. 2002. *Birds of Virginia: Field Guide*. Cambridge, MN: Adventure Publications.

Thieret, John W., William A. Niering, and Nancy C. Olmstead. 2001. *National Audubon Society Field Guide to Wildflowers: Eastern Region*. New York: Alfred A. Knopf.

Tree, Isabella. 2018. *Wilding: Returning Nature to Our Farm*. New York: New York Review Books.

Virginia Department of Forestry. 2016. *Common Native Trees of Virginia: Identification Guide*. Charlottesville, VA.

Whyman, Susan E. 1999. "Sir Ralph Verney: Networks of a Country Gentleman—The Gifts of Venison," in *Sociability and Power in Late-Stuart England: The Cultural Worlds of the Verneys 1660–1720*. New York: Oxford University Press.

Wilson, Edward O. 1992. *The Diversity of Life*. New York: W. W. Norton.

Suggested Reading

Articles

Allombert, Sylvain, Anthony J. Gaston, and Jean-Louis Martin. 2005. A natural experiment on the impact of overabundant deer on songbird populations, *Biological Conservation* 126(1) (November):1–13.

American Philosophical Society. 2023. Locating the transatlantic seed trade in James Madison's garden. Online: amphilsoc.org/museum/exhibitions/historic-meteorological-records-aps/locating-transatlantic-seed-trade-james

Andrews, Betsy. 2023. Bourbon, biodiversity, and the quest to save America's oak forests, *Seven Fifty Daily*. Online: daily.sevenfifty.com/bourbon-biodiversity-and-the-quest-to-save-americas-oak-forests/

Baiser, Benjamin, Julie L. Lockwood, David La Puma, and Myla F. J. Aronson. 2008. A perfect storm: two ecosystem engineers interact to degrade deciduous forests of New Jersey, *Biological Invasions* 10:785–795.

Barden, Lawrence S. 1997. Historic prairies in the Piedmont of North and South Carolina, USA, *Natural Areas Journal* 17(2): 149–152.

Bourg, Norman A., William J. McShea, Valentine Herrmann, and Chad M. Stewart. 2017. Interactive effects of deer exclusion and exotic plant removal on deciduous forest understory communities, *AoB PLANTS* 9(5). Online: academic.oup.com/aobpla/article/9/5/plx046/4107418

Caprara, David. *The Mannahoac Story Alliance: Honoring Virginia's Indigenous Peoples*. YouTube video: youtube.com/watch?v=wdqUdXPgiII

Suggested Reading

Cornwall, Warren. 2018. Common weed killer—believed harmless to animals—may be harming bees worldwide. *Science*. Online: science.org/content/article/common-weed-killer-believed-harmless-animals-may-be-harming-bees-worldwide

Dass, Pawlok, Benjamin Z. Houlton, Yingping Wang, and David Warlind. 2018. Grasslands may be more reliable carbon sinks than forests in California. *Environmental Research Letters* 13(7). Online: iopscience.iop.org/article/10.1088/1748-9326/aacb39/meta

DiTommaso, Antonio, Scott H. Morris, John D. Parker, Caitlin L. Cone, and Anurag A. Agrawal. 2014. Deer browsing delays succession by altering aboveground vegetation and belowground seed banks. *PLoS ONE* 9(3). Online: doi.org/10.1371/journal.pone.0091155

Dorey, Jenna E., Jordan R. Hoffman, Júlia Lins Martino, James C. Lendemer, and Jessica L. Allen. 2019. First record of *Usnea* (Parmeliaceae) growing in New York City in nearly 200 years, *The Journal of the Torrey Botanical Society* 146(1) (February): 69–77.

European Commission, Directorate-General for Environment. *Biodiversity: new IPBES report finds invasive alien species a growing and costly threat worldwide.* Sept. 4, 2023.

Fang, Hui, Conrad C. Labandeira, Yiming Ma, Bingyu Zheng, Dong Ren, Xinli Wei, Jiaxi Liu, and Yongjie Wang. 2020. Lichen mimesis in mid-Mesozoic lacewings, *eLife*. 9.

Fesenmyer, Kurt A., and Norman L. Christensen, Jr. 2010. Reconstructing Holocene fire history in a southern Appalachian forest using soil charcoal, *Ecology* 91(3):662–70.

Suggested Reading

Goode, Erica, 2016. Invasive species aren't always unwanted. *New York Times* (March 1). Online: nytimes.com/2016/03/01/science/invasive-species.html

Halsch, Christopher A., Sarah M. Hoyle, Aimee Code, James A. Fordyce, and Matthew L. Forister. 2022. Milkweed plants bought at nurseries may expose monarch caterpillars to harmful pesticide residues, *Biological Conservation* 273. Online: sciencedirect.com/science/article/pii/S000632072200252X?via%3Dihub

Higgins, Adrian. 2020. Adrian Higgins on renovating your lawn and all things gardening: Q. How to get started, *The Washington Post* (August 27). Online: live.washingtonpost.com/gardening-0827.html

Jirinec, Vitec, Daniel A. Cristol, and Matthias Leu. 2017. Songbird community varies with deer use in a fragmented landscape. *Landscape and Urban Planning* 161:1–9.

Kim, Shi En. August 15, 2023. How swaths of invasive grass made Maui's fires so devastating. *Smithsonian Magazine*. Online: smithsonianmag.com/smart-news/how-swaths-of-invasive-grass-made-mauis-fires-so-devastating-180982729/

Kolakowski, Leszek. 1990. The general theory of not-gardening. *Harper's* (November).

Kranking, Carlyn. 2021. Birds are one line of defense against dreaded spotted lanternflies, *Audubon Magazine*. Online: audubon.org/news/birds-are-one-line-defense-against-dreaded-spotted-lanternflies

Suggested Reading

University of Maryland Extension. 2023, February 15. *Cultivars of Native Plants*. Online: extension.umd.edu/resource/cultivars-native-plants/

Massachusetts Audubon. January 5, 2021. Abundant deer are bad news for understory birds. Online: blogs.massaudubon.org/distractiondisplays/abundant-deer-are-bad-news-for-understory-birds/

McAvoy, Thomas J., Amy L. Snyder, Nels Johnson, Scott M. Salom, and Loke T. Kok. 2012. Road survey of the invasive tree-of-heaven (*Ailanthus altissima*) in Virginia. *Invasive Plant Science and Management* 5(4), 506–512.

Mesnage, Robin, Charles Benbrook, and Michael N. Antoniou. 2019. Insight into the confusion over surfactant co-formulants in glyphosate-based herbicides, *Food and Chemical Toxicology* 128: 137–145

Motard, Eric, Sophie Dusz, Benoît Geslin, Marthe Akpa-Vinceslas, Cécile Hignard, Olivier Babiar, Danielle Clair-Maczulajtys, and Alice Michel-Salzat. 2015. How invasion by *Ailanthus altissima* transforms soil and litter communities in a temperate forest ecosystem. *Biological Invasions* 17: 1817–1832.

Mulhollem, Jeff. 2021, August 3. People ratchet up feelings of burnout in stressed fawns. *Futurity*. Online: futurity.org/fawns-predators-stress-burnout-2606322/

Narango, Desiree L., Douglas W. Tallamy, and Peter P. Marra. 2018. Nonnative plants reduce population growth of an insectivorous bird. *Proceedings of the National Academy of Sciences*, 115(45): 11549–11554

Suggested Reading

National Audubon Society. 2022. Important Bird Areas (IBA), Upper Blue Ridge Mountains. Online: gis.audubon.org/portal/apps/sites/#/nas-hub-site

The Nature Conservancy. 2023. Fire, management and monitoring: using fire in the Allegheny highlands to maintain biological diversity in an ecosystem critical to climate change migration. Online: nature.org/en-us/about-us/where-we-work/united-states/virginia/stories-in-virginia/allegheny-highlands-restoration/

University of Nevada, Reno. 2022. Store-bought milkweed plants can expose monarch caterpillars to harmful pesticides. *Phys.org*. Online: phys.org/news/2022-09-store-bought-milkweed-expose-monarch-caterpillars.html

Oakham, Elisabeth (director). 2022. "Human Worlds," in *The Green Planet* with David Attenborough. Season 1, episode 5. BBC Natural History Unit, UK.

Pavulaan, Harry, and David Wright. 2005. *Celastrina serotina* (Lycaenidae: Polyommatinae): a new butterfly species from the northeastern United States and eastern Canada. *The Taxonomic Report of the International Lepidoptera Survey*, 6(6): 1–19. Online: researchgate.net/publication/270161386_Celastrina_serotina_Lycaenidae_Polyommatinae_a_New_Butterfly_Species_from_the_Northeastern_United_States_and_Eastern_Canada

Quammen, David. 2023. What is wildness? *New York Review of Books* (May 16). Online: nybooks.com/online/2023/05/16/what-is-wildness/

Reichard, Sarah Hayden, and Peter White. 2001. Horticulture as a pathway of invasive plant introductions in the United States: most invasive plants have been introduced for horticultural

Suggested Reading

use by nurseries, botanical gardens, and individuals, *BioScience*, 51(2) (February): 103–113

Relyea, Rick A. 2005. The lethal impact of Roundup on aquatic and terrestrial amphibians. *Ecological Applications*, 15(4): 1118–1124

Rinella, Ken. 2023, October 10. The American buffalo documentary by Ken Burns looks at the slaughter and salvation of bison. *Outside*. Online: outsideonline.com/culture/books-media/american-buffalo-ken-burns/

Roper, Willem. 2020. Western monarchs rapidly declining, *Statista*. Online: statista.com/chart/20716/monarch-butterfly-population-decline/

Rowley, Brent. 2022, Winter. Bison roamed the mountains, too. *Park Science Magazine*. 36:2. Online: nps.gov/subjects/parkscience/issue-winter-2022.htm

Roy, H. E., A. Pauchard, P. Stoett, T. Renard Truong, S. Bacher, B. S. Galil, P. E. Hulme, T. Ikeda, K. V. Sankaran, M. A. McGeoch, L. A. Meyerson, M. A. Nuñez, A. Ordonez, S. J. Rahlao, E. Schwindt, H. Seebens, A. W. Sheppard, and V. Vandvik (eds.). IPBES summary for policymakers of the thematic assessment report on invasive alien species and their control of the Intergovernmental Science-Policy Platform on Biodiversity and Ecosystem Services. 2023. IPBES secretariat, Bonn, Germany. doi.org/10.5281/zenodo.7430692

Shen, Xiaoli, Norman A. Bourg, William J. McShea, and Benjamin L. Turner. 2016. Long-term effects of white-tailed deer exclusion on the invasion of exotic plants: a case study in a mid-Atlantic temperate forest. *PLoS ONE* 11(3). Online: plos.org/plosone/article?id=10.1371/journal.pone.0151825

Suggested Reading

Simberloff, Daniel, Lara Souza, Martín A. Nuñez, M. Noelia Barrios-Garcia, and Windy Bunn. 2012. The natives are restless, but not often and mostly when disturbed, *Ecology*, 93(3) (March): 598–607.

Simberloff, Daniel, and Betsy Von Holle. 1999. Positive interactions of nonindigenous species: invasional meltdown? *Biological Invasions* 1: 21-32.

van Wagtendonk, Jan W. 2007. The history and evolution of wildland fire use. *Fire Ecology* 3: 3-17. Online: doi.org/10.4996/fireecology.0302003

Virginia Native Plant Society. 2015. Resolution regarding deer management. Online: vnps.org/public_html/wp-content/uploads/filebase/VNPS%20Deer%20Management%20Resolution-signed%20with%20header.pdf

Virginia's Prescribed Fire Council. 2020. Beyond the bonfire: a primer on prescribed fire for Virginia's private landowners. Online: dwr.virginia.gov/wp-content/uploads/media/Beyond-the-Bonfire.pdf

Vuocolo, Celia. 2022. Landscape and memory: The bobwhite in virginia, *Quail Forever* (Spring): 24–30

Wickert, Kristen L., Eric S. O'Neal, Donald D. Davis, and Matthew T. Kasson. 2017. Seed production, viability, and reproductive limits of the invasive *Ailanthus altissima* (tree-of-heaven) within invaded environments, *Forests*, 8(226). Online: mdpi.com/1999-4907/8/7/226

Yang, Yi, David Tilman, George Furey, and Clarence Lehman. 2019. Soil carbon sequestration accelerated by restoration

Suggested Reading

of grassland biodiversity, *Nature Communications* 10(718). Online: doi.org/10.1038/s41467-019-08636-w

Yasuoka, J. I., D. Helwig, W. Powell, J. K. Farney, G. F. Sassenrath, and B. C. Pedreira. 2023. Nutrient management strategies to control broomsedge infestation and improve yield and quality of tall fescue hayfields. Southeast Research and Extension Center Agricultural Research. *Kansas Agricultural Experiment Station Research Reports*. 9:2 Online: wildcatdistrict.k-state.edu/agriculture/broomsedge/2023%20 Nutrient%20management%20strategies%20to%20 control%20broomsedge%20infestation%20and%20 improve%20yield%20and%20quality%20hayfields.pdf

Zhang, Luoping, Iemaan Rana, Rachel M. Shaffer, Emanuela Taioli, and Lianne Sheppard. 2019. Exposure to glyphosate-based herbicides and risk for non-Hodgkin lymphoma: a meta-analysis and supporting evidence, *Mutation Research/Reviews in Mutation Research*, 781:186–206

ACKNOWLEDGMENTS

It takes a village to learn about a mountain. When I began this project I knew little about plants, about ecological restoration, or about the resources that were available to me; I came armed mainly with curiosity and stubborn idealism. I could not have completed this book without the help of a long list of people, so many that I fear I'll leave someone out—if I did, that omission doesn't reflect my level of gratitude.

I'm grateful to the community of experts, many more than I can mention here, who were willing to speak with me, to walk with me on the mountain, to answer my many, many questions, and fill in the gaps in my knowledge. Their fields range from forestry to farming to birding to botany to soils to conservation biology and beyond. I want to thank in particular Brian Adams, Matt Dowdy, Kathryn Everett, October Greenfield, Bert Harris, Kadiera Ingram, Charlotte Lorick, Kenner Love, Elizabeth Mizell, Brian Morse, Peter Schoderbek, Erin Shibley, Emily Southgate, Celia Vuocolo, and Natali Walker. I'm grateful to the numerous organizations that shared information and expertise with me, including Blue Ridge Prism, Clifton Institute, Culpeper Soil and Water Conservation District, Friends of the Rappahannock, Mt. Cuba Center, the Natural Resources Conservation Service, Piedmont Environmental

Acknowledgments

Council, the Rappahannock League for Environmental Protection, Smithsonian Conservation Biology Institute's Virginia Working Landscapes, Virginia Cooperative Extension, and Virginia Department of Forestry. I want to especially thank Joyce and Mike Wenger for generously introducing me around and welcoming me to the community.

Thank you to Susan E. Whyman for talking with me about the social history of deer hunting, and for so much more; to Jonathan Wilson of Haverford College for sharing his research on the coevolution of plants and the environment; to Timothy Shively of Virginia Tech for describing his work on a potential biocontrol for the ailanthus; and to Vic Piatt for the entertaining and informative tour of Mt. Cuba Center's native plant gardens.

Every kid who ever turned over a rock to look for potato bugs and grew into an adult who still does that had someone who inspired and encouraged them early on. Thank you to Timothy Riggott, my third grade science teacher, for helping me to understand how everything is connected and for eventually giving me my first writing assignment about the natural world, which included wading around in pond muck that stole one of my galoshes. (It was then I found my writing voice, and it has not changed much since.)

I'm deeply grateful for two writing mentors who passed away in recent years: Daniel Menaker and Kermit Moyer. Their impact on my writing and my professional life has been immeasurable.

I could not have managed the intensive writing and research required for this project without the repeated support of Sheila Pleasants and Dana Jones for fellowships at the Virginia Center for Creative Arts; and Sir Peter Crane and Danielle Eady for residencies at Oak Spring Garden Foundation. The gift of time, solitude, and artistic commiseration has been indispensable for my work. I thank my fellow artists in residence at those places these past few years, including composer Evan Wright, who set an early scene from the book to music for an impromptu live performance; the work of these artists inspired me and helped me to reach higher. Thanks to the Maryland State Arts Council for the partial support of this book through a Creativity Grant; it came at just the right time.

Acknowledgments

There might not have been a book at all if it weren't for Gillian MacKenzie's zeal for this project and her impeccable critical eye; she is the perfect advocate for this book, and I'm lucky to have her on my side. Many thanks as well to Peter Trachtenberg for his early encouragement and for knowing that Gillian would love the book.

I feel fortunate to be working with a talented publishing team whose work centers on books about the natural world and who have been all in on this book from the start. I'm ever grateful to Stacee Gravelle Lawrence for her enthusiasm in bringing me to Timber and her valuable insights on the early pages. It's been a joy to work with my astute editor, Makenna Goodman, whose thoughtful advice made this a better book. Thank you to Kevin McLain for appreciating the humor, to Katlynn Nicolls, Melina Dorrance, and Caroline McCulloch for working to make sure everyone would hear about this book, to Jacoba Lawson for her smooth management of all aspects of production, to Nick Dysinger for his always-friendly efficiency, to copyeditor Laura Whittemore for her discerning eye and for keeping me from embarrassing myself, and to proofreader Callie Stoker-Graham for making sure the book was truly ready. Thank you to the fabulous design team, and especially Vincent James, the visionary cover artist and map artist. Endless gratitude to Lauren Hodapp of Shreve Williams for her boundless energy and creativity, and for going the extra mile on my behalf.

I would need a whole chapter to thank Dave Singleton for his advice, reading, critiques, and moral support at every stage of this project, and Susan Coll for reading many pages and providing much-needed wisdom, honest insights, and reality checks. You both kept me from repeatedly wandering down the garden path (yes, pun). In light of the expert advice I received throughout this project, all errors, stumbles, and regrettable decisions are entirely my own.

To my environmental authors' group, and to the Authors Guild for helping to bring us together, many thanks for sharing ideas and experience, information, support, and camaraderie, in particular Tony Tekaroniake Evans, John Farnsworth, Cynthia Grady, Elizabeth Hilborn, Ralph Lutts,

Acknowledgments

Alan McGowan, Joanna Malaczynski-Moore, Carole Rollins, Christine Wenc, Claire Whitcomb, and Larisa White.

It's important to have people in your life who are willing to cheer you on when you need it and listen to you vent about the big, important things as well as the small, dumb things. Thank you to Carolyn Parkhurst, Leslie Pietrzyk, and Amy Stolls, my perpetually supportive and truthful friends, who can always make me laugh.

Without the love and support of my family, not a word would be written. Thank you to my parents for letting me collect one hundred cicadas and keep horseshoe crabs in the bathtub, and for modeling a style of storytelling that inspired me (rambling and unhurried, with lots of details and, eventually, a punchline). Thank you to David and Eric for your love and your well-developed sense of the absurd, and for being proud to have a mom who writes. And thank you most of all to Bill, always, for believing in me, for picking me up when I'm down, for sticking with me even when I lose track of time watching beetles and talk incessantly about invasive ailanthus trees, and for his patience with everything that comes from living with a writer.

And, finally, thank you to all the living things on the mountaintop, and to the beautiful time-worn mountain itself, for helping me to see.